Human Genetic Biobanks in Asia

This volume investigates human genetic biobanking and its regulation in various Asian countries and areas, including Japan, Mainland China, Taiwan, Hong Kong, India and Indonesia. It sheds light on how cultural, socio-political and economic factors influence the set-up of bioethical regulation for human genetic biobanks and how bioethical sensitivities surrounding biobanks are handled.

Apart from placing discourses of trust in an international perspective, the comparative materials presented in this volume also put into perspective the concepts of genetic theft and exploitation, and genetic wealth and trust. This collection contains case studies of biobanking practices in societies with different needs and welfare levels, and provides insights into government strategies towards genetic resources by examining bioethics as practised at home.

Margaret Sleeboom-Faulkner is Reader in Anthropology at the University of Sussex.

Routledge Contemporary Asia Series

Human Genetic Biobanks in Asia

Politics of trust and scientific advancement

Edited by
Margaret Sleeboom-Faulkner

Routledge
Taylor & Francis Group

LONDON AND NEW YORK

First published 2009
by Routledge
4 Park Square, Milton Park, Abingdon, Oxon OX14 4RN

Simultaneously published in the USA and Canada
by Routledge
605 Third Avenue, New York, NY 10017

Routledge is an imprint of the Taylor & Francis Group, an informa business

First issued in paperbak 2012

Typeset in Times New Roman by
Book Now Ltd, London

British Library Cataloguing in Publication Data
A catalogue record for this book is available from the British Library

Library of Congress Cataloging in Publication Data
Human genetic biobanks in Asia: politics of trust and scientific
advancement/edited by Margaret Sleeboom-Faulkner.
     p. ; cm. — (Routledge contemporary Asia series; 12)
Includes bibliographical references.
1. Human gene libraries—Asia. 2. Biobanks—Asia. 3. Human
genetics—Government policy—Asia. I. Sleeboom-Faulkner,
Margaret, 1961– II. Series.

[DNLM: 1. Biological Specimen Banks—ethics—Asia. 2. Biological
Specimen Banks—legislation & jurisprudence—Asia. 3. Cultural
Characteristics—Asia. 4. Databases, Nucleic Acid—ethics—Asia.
5. Databases, Nucleic Acid—legislation & jurisprudence—Asia.
6. Socioeconomic Factors—Asia. QU 23 H918 2009]

QH442.4.H86 2009
174′.957095—dc22   2008027933

ISBN13: 978–0–710-31341–6 (hbk)
ISBN13: 978–0–415-54141–1 (pbk)
ISBN13: 978–0–203–88460–7 (ebk)

Contents

PART II
Biobanking and problems of inequality

Illustrations

Tables

Figures

Contributors

Ho, Connie Department of Psychology, University of Hong Kong, Hong Kong.

Kumar, Nandini K. Indian Council of Medical Research, New Delhi, India.

Liu, Hung-En Department of Law, National Taipei University, Taipei, Taiwan.

Masui, Tohru JCRB Cell Bank, Division of Bioresources, National Institute of Biomedical Innovation, Osaka, Japan.

Patra, Prasanna Kumar International Institute for Asian Studies (IIAS), Leiden, The Netherlands.

Porter, Gerard AHRC Research Centre for Studies in Intellectual Property and Technology Law, School of Law, University of Edinburgh, Edinburgh, United Kingdom.

Sleeboom-Faulkner, Margaret Department of Anthropology, University of Sussex, Brighton, United Kingdom.

Sung, Wen-Ching Department of Anthropology, University of Toronto, Toronto, Canada.

Tai, Terence Hua Department of Philosophy, National Chengchi University, Taipei, Taiwan.

Waye, Mary Miu Yee Department of Biochemistry, Croucher Foundation of Human Genomics, Chinese University of Hong Kong, Hong Kong.

Zhang, Xinqing Center for Bioethics, Peking Union Medical College, Beijing, People's Republic of China.

Acknowledgements

This collection of case studies is the result of collaborative efforts of an international group of scientists and thinkers studying the impact of biobanking and the establishment of genetic databases on society. They have exerted great efforts, both linguistically and in the area of theorising the socio-economic and cultural dimensions of biobanking.

This volume would not have come about without the support and encouragement of the International Institute for Asian Studies in Leiden (IIAS) and the Genomics Initiative of the Netherlands Organisation for Scientific Research (NWO). These organisations shouldered a large part of the coordination and expenses of the meetings on *Biobanking in Asia*, held in Shanghai in August 2004, out of which this volume has grown. Here I would also like to acknowledge the generosity of *The Asian Social Science Journal* (ASSJ), who gave permission to republish an altered version of Chapter 6. Finally, I would like to express my gratitude to the book's English copy-editor, Lee Bowers, whose enormous efforts are very much appreciated by all of the chapter contributors.

Introduction

1 Human genetic biobanking in Asia

Issues of trust, wealth and ambition

Margaret Sleeboom-Faulkner

This volume aims to provide a background to biobanking projects and discussions regarding their establishment in Asia by providing socio-economic and political contexts, and a variety of views on the institutional context in which biobanks are set up and genetic sampling takes place. These views include those of scientists and regulators on bioethical issues and those of vulnerable communities and individuals. Apart from placing discourses of trust in an international perspective, the comparative materials presented in this volume also put into perspective the concepts of genetic theft and exploitation, and genetic wealth and trust. This collection contains case studies of biobanking practices in societies with different needs and welfare levels, and it tries to provide insights into government strategies towards genetic resources by examining bioethics as practised at home.

Unfortunately, many bioethics discussions have been held as if research and biobanking take place in a socio-economic vacuum (Martin and Kaye 2006: 96). And in a medical context, bioethical issues are often just related to the informed consent procedure of donors if it concerns the database only (Williams and Schroeder 2004). When discussions do take socio-economic factors into account, they often take for granted a European socio-economic and cultural context of systems of welfare and healthcare provision and democratic debate, unaware of the similarities and differences within Asia and between Europe and Asia. And when global factors are taken into account, culture tends to be reduced to religion or stereotyped as Oriental (Sleeboom 2004). For instance, when at a public meeting on genomics and society in The Hague, in 2002, the author put forward as an example of dialogue between scientists and the public the efforts made by 'Genome Square' in Japan, it was waved away as 'We all know about privacy and authority in Japan'. Considering the prevalence of hackneyed images about Japan, it is easily assumed that discussions on biobanking and privacy in Japan are a façade for authoritarian action, while those held in the UK somehow are taken more seriously. It may come as a surprise to many Europeans, then, that privacy in Japan constitutes a greater stumbling block to successful biobanking than it does in Europe, mainly due to a fear of an increase in instances of discrimination (Ch. 3). Similarly, biobanking in China would facilitate socialist state control over its citizens, due to the nation's paternalistic and authoritarian ideologies, such as Confucianism and socialism. Such assumptions, however, are ungrounded when not backed up by concrete empirical evidence and blind us to other, more plausible views.[1]

This volume attempts to discuss a number of features of human genetic biobanking and its regulation in various countries in Asia, including Japan, Mainland China, Taiwan, Hong Kong, India and Indonesia. It aims to shed light on how cultural, socio-political and economic factors influence the set-up of bioethical regulation for human genetic biobanks and how bioethical sensitivities surrounding biobanks are handled. First, it introduces the notion of biobank and its uses. This is followed by a brief discussion on the organisation and forms of genetic biobanks, genetic sampling practices in Asia, and an analysis of the criteria used to define research populations. In short, this introduction hopes to indicate some of the features and problems of genetic biobanking in Asia related to the main themes running through the chapters of this book: trust, wealth and ambition.

The nature of the relation between medical research and society is changing in fundamental ways. The focus of medical research has shifted away from the study of individual bodies as physical entities to medical studies based on networks of data, information and archival systems, such as biobanks. A shift is taking place, then, from the hospital as site of biomedical research to the laboratory. At the same time, the management of public and individual healthcare has come to depend on technologically advanced systems for maintaining healthcare data, and blood and tissues in biobanks, requiring new methods of data collection, maintenance and management. A seemingly contradictory process is taking place in the management of healthcare data and tissues. On the one hand, body tissues and data are processed, stored and standardised so as to become applicable to broad categories of potential patient populations. On the other hand, data and tissues are re-assembled to fit the specific individual in tailor-made medicine (Waldby and Mitchell 2006). In short, biobanks are central to the organisation, maintenance and manipulation of data and tissues from laboratory and hospitals.

A biobank can be used for the collection, storage, maintenance and application of genetic samples and information. The information in human genetic biobanks affects the organisation of social life and medical possibilities in many ways. For instance, it facilitates the expansion of research efforts in the fields of forensic science and epidemiology. Thus, the police use forensic genetic databanks to match with DNA found at the location of a crime, and genetic biobanks can be used in studies aimed at determining the relationship between disease and the genetic make-up of individuals and groups. This relation, again, can be matched with ethnic background, living environment and socio-cultural group membership.

The creation of genetic biobanks also makes possible the collection and management of personal information, whose disclosure and control has far-reaching social, political and financial consequences. The disclosure of the genetic make-up of an individual with a serious genetic defect, for instance, may have considerable consequences for his or her family relations, employment, insurance possibilities and life expectation. When a genetic defect is linked to a group of people or lifestyle, the consequences may be even more far-reaching and lead to socio-genetic discrimination. Human genetic biobanking, then, is of great concern to society, and has been the object of discussion in the social sciences in various countries, though mainly concentrated in European countries and in North America (Knoppers 2003; Sándor 2003; Sutrop 2004).

These discussions often take the form of a public debate. Such debates partly are meant to educate the public about the use of biobanks in science and in social life. Such public debate is of particular importance to regulatory authorities, as the success of biobanks depends largely on the cooperation of the public. For this reason the creation and management of biobanks calls for sensitive policies regarding the regulation of biobanking activities. Biobanking and regulation, however, take different shapes in different parts of the world, largely due to the existence of diverging systems of socio-political and economic organisation, different functions of and attitudes towards science, and different ways in which governments relate to their populations.

Categorising human genetic biobanks

A biobank is essentially a database that stores patient and population samples, including tissue samples, serum and genetic information, for the purpose of medical research. There are various terms used for collections of biological materials and the genetic information derived from them. The UK Biobank is regarded as a population-based genetic database; the Estonian Genome Project considers itself as a gene-based data collection; the Latvian Project calls its DNA collection a genetic population bank; and COGENE (Coordination of Genome Research Across Europe) regards biobanks as cohort studies, which in this case are essentially long-term comparative studies of human genetic groups (also see Bhardwaj 2007).

The biobanks referred to in this volume concern the activities around systematically organised human DNA, set up mainly for the study, the use and the preservation of genetic information,[2] including biobanks for population studies, migration and genomics research. Biobanks do not only have different purposes; they also differ through the ways in which they are regulated, the methods they use for organising data, and the ways in which they are maintained. All of them, however, are interested in how genes interact with their environment, including, climate, pollution, life-style and psychological factors relevant to human behaviour. Some efforts have been made to define genetic biobanks with some sensitivity of DNA-banking activities to individuals in society. And finally, it is clear from definitions of genetic databases and biobanks as collections of data and information (House of Lords 2001; WHO 2002), that genetic data, as well as genetic samples, are main intellectual resources for bioinformatics and the study of gene–environment interaction (Bhardwaj 2007: 134).

To gain some insight into the great variety of databases and biobanks a brief discussion on the criteria used to distinguish between them may be useful. As pointed out, a distinction can be made between biobanks that work with biological samples and those that contain data of samples that have been processed, analysed and electronically stored. This distinction is especially important to the distribution activities of biobanks, some of which distribute the actual samples (such as Biobank UK) and others that only distribute data (Biobank Japan) (Ch. 4). The databases are also distinguished according to the variety of ways in which they store data, such as can be found in specialised genomic databases, mutation databases, and as databases for protein sequences and nucleic acid sequences.

An important criterion is politico-geographical. A relatively recent, but nowadays common way of referring to population-based data collections is by focusing on the national, regional or international nature of their sampling targets. Large-scale population-based genetic databases are used for screening programmes, genetic epidemiological studies, genetic association studies, and genealogies.[3] They also form the basis for applications in pharmacogenomic and nutrigenomic studies and tailor-made medicine. Although transnational companies often initiate industrialised applications, the initial gathering usually leaves a geographical footprint on the biobank. Examples of national databases are the Norwegian biobank,[4] the Icelandic Health Sector Database,[5] the Estonian Genome Project,[6] the Swedish Karolinska Institutet-Biobank,[7] Biobank UK,[8] the Finland GenomEUtwin Consortium,[9] Biobank Japan (Ch. 4), Taiwan Biobank (Ch. 5) and the Singapore Tissue Network.[10] Examples of regional studies are CARTaGENE, a database that targets the Quebec population for medical, sociological and personal data,[11] the Bristol-based ALSPAC project,[12] and the North Cumbria Community Genetics Project (NCCGP)[13] in the UK, the Guangzhou Biobank in the South of China,[14] and the 'world's largest gene bank for ethnic minorities in Kunming in Southwest China.[15]

Other biobanking projects combine international efforts, funding, genetic materials and data in population-based genetics. The international haplotype map (HapMap) project is a collaboration among scientists and funding agencies from Japan, the UK, Canada, China, Nigeria, and the USA, set up to assemble DNA polymorphisms for the discovery of SNPs that affect gene function, facilitating studies related to human genetic variation to disease risk.[16] International biobanking projects have been initiated to facilitate the discovery of disease genes for complex traits. However, some concerns about who will have access to the data, which groups will be able to turn these data into profits, and which genetic diseases will be the focus of research led some Asian nations to set up their own haplotyping initiative.[17] Another example is the Howard University initiative of the Genome Research in African Diaspora (GRAD) that aims to set up a DNA repository of black people to study diseases prevalent among African-Americans in particular.[18]

Other distinctions between biobanks are between commercial and non-profit and public and private biobanks, although they can be confusing and unhelpful. As for the distinction between commercial and non-profit, large-scale genetic biobanks usually rely for a large part on the investment of the pharmaceutical industry and need the state sector for logistic, regulatory and institutional support. As for the distinction between private and public biobanks, the term public also has become applicable to private biobank initiatives, when it refers to the openness of the distribution function of the data collection (Ch. 4). The public image of the biobank is also of importance to the bank's need to solicit cooperation from the public. Privately funded by deCODE Genetics, the Icelandic Biobank, intended to collect and analyze DNA samples for commercial purposes, funding research that seeks to identify the genetic causes of common diseases in order to improve their treatment and diagnosis.[19] The Canadian-initiated Public Population Project in Genomics (P3G) tries to improve the accessibility of population-based projects to a global public. It aims to stimulate international collaborative research and the regulation of research methods between population-based genetic databases. It also hopes to harmonise criteria used by members of population databases from various disciplinary backgrounds (Bhardwaj 2007: 135).[20]

It is also possible to categorise genetic biobanks according to the kind of human targets they investigate. The Norwegian Mother and Child Cohort Study collects biological samples and health and exposure data from pregnant women, children and fathers,[21] while the GenomeEUtwin is an international cohort study involving Danish, Finnish, Italian, Dutch and Swedish twins. Other collections focus on specific *ethnic minorities*, such as the ethnic minority genebank in Kunming mentioned above, and the Centre for Cellular and Molecular Biology in Hyderabad, which has collected 9000 samples from 130 endogamous populations in India.[22] Various other criteria, not discussed here, could be used to characterise biobanks, such as size, commercial or non-profit, and the biobank's timeframe (Ch. 4).

For now, it should be clear that any generalisation about the usefulness and social threat of genetic biobanks might be problematic. Another approach to biobanks observes the ways in which research populations are defined in units or groups of people. Crucial to the success of research is the question of what criteria to use for defining research populations when analysing research problems. How do biobankers decide which data to sample, and what criteria do they use when planning their research; that is, before they know the genetic make-up of populations? This question is interesting especially in relation to indigenous populations, as it is hoped that genetic populations relatively untainted by modern medicine and lifestyles can provide useful material for the research of affluence-related diseases especially prevalent in welfare societies, such as forms of diabetes, and heart and lung diseases. The following section discusses issues related to the search for useful genetic populations in developing countries; the next section focuses on criteria used to define research populations; and the section after that concentrates on issues of trust and privacy in general.

International biobanking and genetic sampling in Asia: ethnicity, nationality and bioethical regulation

The collection and storing activities of DNA-banking serve various areas of research. As part of the Human Genome Project (HGP), which was initiated in 1991, these forms of data collection served the 'completion' of the human genome map in 2001 (Lander *et al.* 2001; Venter *et al.* 2001). While some scientists study the 'average' or 'standard' genome, others study variations within and between populations. An increasing number of projects now focus more directly on human variation and diversity by sequencing the DNA of selected, and supposedly more genetically 'pure', 'indigenous populations'.[23] Detecting genetic variation among such populations becomes easier if there is a generic sequence for comparison. And as the HGP has 'completed' the map of the 'average' human genome, there has been increased interest in sampling indigenous populations in order to study human diversity.

Population geneticists hope that identifying differences in genetic sequences between peoples will help determine what makes groups of people different from other groups. Of course, other factors also contribute to population differences, such as culture, language, environment and lived experience. Although in some population studies these factors have been hard to estimate (e.g. Majumder 2000), it is hoped that the identification of genetic difference between ethnic groups (such as a different gene sequence that causes immunity to a disease) will have human medical, cultural and historical and economic value. However, this perceived value in the 1990s also resulted in a 'gold rush' where universities, governments, corporations and private researchers are seeking to identify human genetic variation (Marshall 1997: 565).

Another international effort, the Human Genome Diversity Project (HGDP), started earlier than the HGP and involved genetic sampling among indigenous populations. According to Luca Cavalli-Sforza, its initiator, the HGDP served to avoid the irreversible loss of precious genetic information (Cavalli-Sforza *et al.* 1991: 490–1). Referring to indigenous populations as 'isolates of historic interest (IHIs)', the HGDP planned to immortalise the DNA of disappearing populations for future study (cf. Human Genome Organisation [HUGO] Committee 1995). In this context, one aim of the creation of gene banks for scientists was to reconstruct the history of the world's populations by studying genetic variation to determine patterns of human migration.

However, the initial conceptualisation of the HGDP has been widely criticised for its consideration of indigenous peoples as mere research subjects, with little regard for their continued livelihood. Various ethnic groups have protested against population genetics, which they claim has been conducted without prior consultation and without consulting the individuals and communities involved (cf. Knoppers 2003; Peterson 2001; Pilnick 2002; Sándor 2003;

Sutrop 2004). Because the HGDP was planning to collect blood samples, some groups called the HGDP the 'Vampire Project' (Lock 1994: 603–6), while others were angry because they believed that they were possible target populations even though no community representatives had been contacted about the project.[24]

In some developing countries with great ethnic diversity and little access to healthcare, such as India, China and Indonesia, attempts have been made to sample the populations and map genetic resources cheaply. Such populations are targeted because they are thought to be genetically isolated, and in some cases have well-documented genealogical histories, making them valuable for tracing inherited diseases. In Asian developing countries such as India, Indonesia and China, tribal and isolated communities, which are the likely subjects for diversity studies, live in remote areas, are often illiterate and suffer from nutritional and parasitic disorders. Here, it may be important to distinguish between studies that are aimed at improving the healthcare of the local community (Ch. 8) and those that aim at sampling isolated communities to study the diseases of the wealthy, such as diabetes and coronary heart disease. For instance, the Indian Kondh tribe in Orissa were studied for sickle cell disease, and Negrito tribes of the Andaman Islands were evaluated to find the cause for their demographic decline. According to Ishwar Verma (2000), these studies observe procedures of informed consent from both individual and community perspectives, provide compensation for expenses incurred, and are accompanied by the provision of medical care and reproductive advice during the study.

In the 1990s, India had become the hunting ground for genomics, as Muslim groups in the north and consanguineous marriage in the south are an ideal setting to search for new genes by the technique of 'homozygosity mapping' (Verma 1997). Since laboratories with facilities for identifying new genes are very few in India, samples of these families were sent to scientists in Western counties. A similar situation obtained in the PRC, where a diplomatic dispute between the PRC and the USA came about after the sampling practices of Harvard researcher Xu Xiping amongst the poor in isolated mountainous areas had been questioned by the *Washington Post* (Pomfret and Nelson 2000; Sleeboom 2005). In India and the PRC, draft legislation was drawn up in 1997 and 1998, respectively, to protect biological resources from being exploited by foreigners without informed consent and sharing its benefits with the local population (Jayaraman 1997). In both cases it meant that the transfer of commercially valuable research material to foreign nationals without official approval became illegal (ICMR 2000/6). By contrast, a country such as Indonesia, which suffered from political instability and lacked effective protective measures to keep its genetic resources at home, remained a genetic safe haven for internationally oriented scientists (Ch. 6).

It should be clear from these accounts that awareness of the financial and socio-political consequences of biobanking studies are of great importance to the long-term success of the activities of genetic population banks. Historical awareness, however, may be just as important to the political acceptance of sampling practices. In the Southern Pacific state of Tonga, church and human rights groups opposed plans of the Melbourne-based company Autogen to establish a database of genetic information of Tonga's population. In November 2000, Autogen had informed the Australian Stock Exchange that it had signed an agreement with the Tongan Minister of Health, Dr Viliami Tangi (who denied it), to establish a research project aimed at 'identifying genes that cause common diseases using the unique population resources in the Kingdom of Tonga' (Burton 2002). According to Greg Collier, Autogen's chief scientific officer, Autogen's Tonga project aimed to collect 'information on common diseases of "affluence" – including cardiovascular disease, hypertension and certain cancers' (Nicholson 2000). In return for access to these samples and data, Autogen would provide annual research funding to Tonga's Ministry of Health in addition to paying net royalties on revenues generated from any discoveries that are commercialised. However, the director of the Tonga Human Rights and Democracy Movement, Lopeti Senituli, declared at the Australasian Bioethics Association Conference in Adelaide: 'We really cannot afford to go back to the frontier days when it was open season on all things indigenous to the Pacific Islands' (Senituli and Boyes 2002). The overall perception was that foreign investment serves foreign interests, causing general apprehension and suspicion towards foreign company researchers. Senituli claimed that the opposition was mainly set off by the secrecy surrounding the deal: 'What is involved is the sanctified blood of human beings and not the genetic make-up of our pigs (with all due respect), so there should have been prior public discussions before the Minister signed' (Radio Australia 2000). In 2001, sentiments had risen to such heights that Tonga's association of churches decided to refuse any cooperation with research linked to biotechnology.

Especially where there is no effective national legislation or a regional mechanism to regulate the transfer of samples and data and to regulate biological genetic research in developing countries such as Tonga and Indonesia (Ch. 6), sooner or later critics are bound to demand detailed information on how and where their samples and data will be used and stored. In countries that decided to strictly regulate the transport of biomaterials, such uncontrolled *international* exploitation of genetic materials seems to have subsided. This does not mean, however, that the *internal* exploitation of genetic materials came to a halt. The situation is complex. For although the American National Institute of Health (NIH) declared that the HGDP 'was never carried out' due to 'concerns that the Project would exploit vulnerable individuals and populations and the Project's potential intrusion into cultural beliefs about population origins' (National Human Genome Research Institute 2002), several Asian countries, such as India, Indonesia and China, have been interested in setting up their own regionalised 'HGDPs'. And the biobanking projects in these countries in many cases are or may become subject to the same bioethical scrutinisation as did the projects that involved foreign initiators (see Ch. 10; Sleeboom 2005).

While in the West criticism of the HGDP had been growing in the 1990s, plans for a so-called international haplotype map (the HapMap) were gaining support. It was hoped that commercial and public efforts to assemble DNA polymorphisms would contribute to the discovery of SNPs that affect gene function. Most successes in finding genes that contribute to disease risk have been for so-called 'Mendelian' disorders caused by single genes, such as cystic fibrosis. To locate such genes, researchers perform linkage analysis on families, which requires 300–500 highly informative genetic markers spanning the entire human genome. Locating the genes contributing to the risk of common diseases such as diabetes, heart disease and psychiatric disorders is considered much more difficult, as multiple genes and the environment contribute to these phenotypes (cf. Collins *et al.* 1998). Instead of linkage analysis, a less Herculean task would apply association analysis onto many affected and unaffected individuals. The resulting search for DNA polymorphisms aims to discover the required hundreds of thousands of variants spread over the entire genome. In short, it is hoped that comparing groups of individuals affected by a disease with groups of unaffected controls will lead to the discovery of these DNA-sequence variants that will provide new insights into the pathogenesis, diagnosis and treatment of human disease (cf. National Human Genome Research Institute 2002; International HapMap Consortium 2003).

As part of this global initiative to create the HapMap (participants are the USA, the UK, Japan, China, Canada) and commercial haplotyping, genetic sampling is supposed to be exclusively used for the advancement of medical knowledge, focusing on the differences between the genetic make-up of different individuals only in the area of medicine. This claim, however, is open to dispute, as patterns of variations between individuals are still generalised over entire populations for political reasons, out of prejudice and simply out of habit. As also shown by genetic research on individuals of Ashkenazi Jewish descent and African-Americans, concerns are raised about the augmentation of already existing prejudice and genetic pretexts for discrimination in situations of conflicting political and socio-economic interests.[25] Moreover, concerns exist about who will have access to the data, which groups will be able to turn these data into profits, and which genetic diseases will be the focus of research.

Sample populations for biobanks: criteria selection

Other questions concerned the ways in which populations were defined in the first place. Would haplotyping privilege some group definitions over others? What would be the effect of the creation of new socio-genetic boundaries between populations? A number of factors are relevant to the delineation of populations for genetic sampling. In this volume, socio-political factors constitute the hub of discussion, but it is also important to draw attention to the linguistic and environmental criteria used to define sampling groups, assuming a correspondence between linguistic and genetic development or environmental isolation and genetic development. The methods population geneticists use include genealogical records and linguistics. The use of these methods, per definition, leads to familiar research problems in history and culture, which do not assume languages to be static: languages migrate and change not necessarily parallel with the biological make-up of genetic groups (Cavalli-Sforza 2000); and genealogical records are often faulty, or manipulated for religious, personal and political ends. It is a fact of life that the 'family' is not just a biological concept, but also a phenomenon understood through changing cultural meanings (cf. Clarke and Parsons 1997). For this reason it is difficult to define a stable research population, not just for anthropologists, but also for geneticists.

It is also difficult to define communities in an unambiguous way, including linguistic communities. Although in many cases language spreads through migration (horizontal transfer) and is paralleled by genetic transfer, this is not always the case. After conquest, for instance, language change may take place unaccompanied by substantial genetic admixture. Conversely, genetic interchanges between groups may take place over linguistic boundaries. Thus, genetic profiles of twenty-eight populations sampled in the PRC supported the distinction between southern and northern populations. Migration across linguistic boundaries, i.e. the transgression of language families, changes the genetic make-up of linguistically defined populations, reflecting substantial gene flow between populations (Chu 1998: 11763).

Other definitions of populations used in genetics are based on concepts of race or nation, and on the specific markers used in population genetics. The concept of race includes Caucasian, Negroid and Mongoloid races, which forensic geneticists and epidemiologists categorise as 'population substructures',[26] while national, political and regional boundary markers mark people as Chinese, Taiwanese, aborigine, Mainland Chinese and other categories. Thus, the Chinese people from Taiwan, from a Mainland Chinese point of view, are considered to have an impact on the genetic 'homogeneity' of the Chinese population in the PRC at large, as they contribute to the genetic make-up of China as a whole. In other words, if the genetic make-up of the inhabitants of Taiwan can be shown to resemble those of groups in Mainland China, then Taiwan is said to be part of China. For example, an article in the *Beijing Times* entitled 'DNA Analysis Reveals Taiwanese Have Ancestors on Mainland' announced that DNA analysis by Li Hui, from the Life Sciences Institute from Shanghai Fudan University, had revealed that at least four ethnic groups in Taiwan (the Amei, Yatai, Bunong and Paiwan) share the same chromosome with the Li ethnic people now living on Hainan Island, and they are all descended from the ancient Baiyue people in East China. This shows, according to the report, many Taiwanese originated in the mainland and cross-straits exchanges began 6000 years ago.[27] As lucidly described by Amâde M'charek (2000), national boundaries are seen as prohibiting conditions for what biological theory regards as 'random mating' between members of different populations and as enhancing random mating within the population. In this case, Mainland Chinese official theories say that this boundary includes Taiwan, whereas official Taiwanese theories only include *parts* of Mainland China, while emphasising Taiwan's Austronesian genetic and historical ties.

Another concept of population presumes that populations may be tied to specific markers.[28] Depending on which markers are used, the population unit may be produced differently. Depending on markers, alleles may be equally spread over the whole world; they may be clustered in specific patterns. Here, the population is a product of genetic markers. Thus, the clustering of populations into units of research depends on the type of markers used. For instance, if we use the disease marker for thalassaemia, then we find that southern Chinese people and Mediterranean people are genetically closer than are northern and southern Chinese people.[29] Similarly, haplotyping seventy-four Chinese geographical populations, Tongmao Zhoa and Tsung Dao Lee were surprised to find that three Chinese populations (Uygur, Kazak and Dongxiang) were clustered together with populations for Finnish, English, Iranian and Uralian of the USSR (Zhao and Tsung 1989). Depending on migration routes used at some point in the past, it is not a miracle if certain genetic markers reveal connections between populations previously considered far-flung.

The difficulty with defining a research population on the basis of genetic particularity seems obvious: we do not know its suitability unless we first sample the population. And even if we find a shared genetic trait, we still do not know if groups elsewhere share the same trait until we have sampled all people. Neither do we know whether the trait is relevant as a criterion of distinction in the case concerned. The research experience of Harvard epidemiologist Xu Xiping in southwest China, for instance, shows that this problem lies at the heart of selecting a population to sample. Xu studied asthma in the belief that the genetic make-up of the inhabitants in the remote mountainous regions of Anqing would show the genes responsible for syndromes associated with asthma in what he considered to be an isolated part of the world. But as he had overestimated the genetic homogeneity of his research population in Anqing and underestimated its complexities vis-à-vis the environment, he had to change his research focus (Sleeboom 2005) In other words, conventional political, social, cultural, linguistic and geographical knowledge all provide clues to the genetic make-up of peoples, which sometimes provide genetic samplers with very shaky ground to build upon.

Promise, privacy and trust in biobanking

Although major research successes have been booked in the case of Mendelian, single-gene disorders (and disorders caused by relatively few genes), such as cystic fibrosis, thalassaemia and Huntington's disease, such diseases are relatively rare. The search for the genetic basis of the complex traits of common multifactorial diseases, however, has been disappointing (Lock *et al.* 2005). Multiple genes or environmental factors lead to the complex diseases and each gene (or environmental factor) may have only a small effect on the phenotype or the way in which the disease is expressed. Furthermore, various relationships between gene and environment are responsible for disease occurrence, including gene–gene interaction, gene–environmental factor interaction, one gene causing multiple diseases, and one disease caused by different disease genes (genetic heterogeneity). The complex, non-Mendelian category of diseases includes diabetes, cardiovascular diseases, obesity, and manic depression, as well as schizophrenia. And it is precisely this broadest category of genetic disorders in which genetic studies have not met with the anticipated success (Glazier *et al.* 2002).

Nevertheless, it is clear that systematic knowledge about the genetic make-up of people is of great value to the understanding of the differences in the human susceptibility to diseases, drug responses and the complex interaction of genetic and environmental factors in the production of particular phenotypes. Such knowledge is essential, for instance, when administering medicine to individuals from particular genetic population groups and when conducting human experiments (Berg 2004). Furthermore, the value of biobanking in gene–environment interactions has been demonstrated in many case studies, of which the SARS patient bank study of the susceptibility to SARS coronavirus infection is just one example (Ch. 5) (Mak *et al.* 2006).

The possibilities of identifying new drug targets and discovering drugs through DNA technology have led to expectations of high revenues and the anticipation of increased financial gain in science, universities and companies. Although providing clues to new drug discovery, according to critics, biobanks and databases have become valued beyond their scientific potential. While a promising field, expectations that were raised about the possibilities of tailor-made medicine and applications in public health have not been fulfilled. This, however, is also related to the availability of clear regulatory frameworks and economic incentives and demand from clinical users (Hedgecoe 2005; Hopkins *et al.* 2006; Martin 2006).

The wishes of biobankers to elicit the cooperation of the public in donating samples, combined with concerns from the public about enterprises exploiting donated DNA samples and worries about the level of protection of data collections, led to calls for the regulation of biobanks. In other words, it was hoped that regulation would prevent discrimination, remove personal and public anxiety and encourage public participation in important research. In Europe, the response to calls for bioethical regulation is reflected in the regulatory efforts exerted in the creation of the national biobanks, such as those in the UK, Iceland and Estonia. The growing scientific demand, within both the private and public sectors, for the creation of genetic databases has led to debates on questions relating to the control, use and ownership of personal and medical data.

Discussions on these questions initially focused on the controversial plan in Iceland to create the Health Sector Database, as well as on the creation of similar large-scale databases in Estonia and the UK (Rose 2001). In the press, discussions on biobanking have concentrated on future benefits for biomedical research, the advance of genetics and historical knowledge and the benefits for society in general and the individual in particular (cf. McKie 2000). Doubts about the misuse of biobanks especially emphasise issues of privacy (e.g. the anonymisation of data), autonomy (informed consent) and trust. In this volume, the issues of privacy and autonomy, together with those linked to the protection of information and benefit sharing, are closely associated with the category of trust (Enserink 1998: 859). The reason for this is that tissue donation is usually made on the trust that tissues and information are not used against the interest of the donor, that the information supplied to the donor is as complete and accurate as possible, and that the receiving institutions do not pass on information or tissue without the consent of the donor. Furthermore, other more abstract notions of trust play a crucial role in debates about biobanks, which give it a more ephemeral dimension. Thus, people's trust in their government to protect them or in their society to respect them seem to be closely related to their willingness to participate in a biobanking project (Ch. 5).

An important source of worry in relatively wealthy societies equipped with computerised healthcare systems is the linkage of healthcare records with genetic data (and with genealogical records, if available). The automatisation of paper healthcare records requires the availability of IT and the standardisation of healthcare data and records, which may be costly and time-consuming. The maintenance of vast networks of genetic information, including family medical histories, genealogical data and lifestyle information, involves the use of bioinformatics, which combines genomics and advanced computing technology. Electronic health records in principle are easily copied and communicated to other stakeholders, entailing new risk of the distribution of sensitive data against the will or interests of the individuals concerned. Bioinformation is therefore central to controversies regarding the control and ownership of genetic information, including the patenting of DNA sequences, the reuse of patient records commercially, and the commodification of the genetic samples in the HGDP. The donation of bioinformation is the subject of intensive debate and is discussed in Part I of this volume, that is, the chapters relating to the relatively wealthy Asian societies with healthcare systems.

Chapter overview

This volume is divided in three sections, comprising this introductory chapter and two parts. Part I concerns discussions on biobanks in relatively wealthy modern nation-states, including Japan, Hong Kong and Taiwan, while Part II discusses biobanking in developing countries, such as India, Mainland China and Indonesia.

Chapter 2, by Hung-En Liu and Terence Hua Tai, describes the pilot project of the Taiwanese Biobank Project and its concerns in the area of trust, commercial involvement and benefit sharing. The authors believe that, although support from the pharmaceutical industry seems to be inevitable, a tension between commercial involvement and altruistic donation by the public may damage public trust and support for the biobank in the long run. The authors argue, therefore, that great care should be taken when dealing with the recruitment of donors, especially when competing interests are involved, such as those between physicians and patients. The examples of sampling practices in

Taiwanese hospitals, among Taiwanese aborigines and by the Academic Sinica provide a clear understanding of the difficulties that occur in biomedical research practice in this regard. The linkage between records of healthcare data and the biobank is also a main source of contention. According to the authors, benefit sharing may be one way of counterbalancing the possibility of exposure to discrimination and the cooperation of the donors against the profit and other benefits to companies and researchers. Using examples of the experiences of a number of biobanks in the world, the authors acquire a clear picture of the problems and possible solutions faced by the pilot study of the Taiwanese Biobank.

Comparing the regulatory frameworks of UK Biobank and Biobank Japan, in Chapter 3, Gerard Porter focuses on the theme of acquiring public trust in support of biomedical research. In their attempts to set up biobanking projects, both the UK and Japan had to overcome mistrust in the light of medical scandals, but Japan also has to deal with issues of paternalism in patient–physician relations, mistrust in politics, and a hypersensitivity regarding privacy due to a deeply entrenched fear of social discrimination. Although detailed guidelines have been developed with regards to the duty to obtain full informed consent and to protect the privacy of research participants, problems remain unsolved regarding the power invested in individual Ethics Review Committees, the conditions under which samples may be linked to research subjects and medical records, and the status of the administrative guidelines. Critics refer to the UK Biobank efforts to win over the public, when discussing the lack of public engagement regarding biobanks and their benefits in Japan.

In Chapter 4, Tohru Masui from the National Institute of Biomedical Innovation, describes a development in biomedicine of an increasingly closer link between clinical and laboratory research. Biobanks reflect a need for genetic data that results from this development. For instance, biobanking is expected to yield genetic insights useful for the following generations and it is hoped that it will make redundant some human medical trials as, for instance, data on the interaction between drugs and DNA become available. Genomic research has generated several biobanks in Japan. After providing us with a typology of biobanks and an overview of the regulation of genomics in Japan, Masui describes the activities, protocols and bioethical regulation of two kinds of biobanks: the Pharma SNP Consortium, and Biobank Japan. In this chapter, Masui discusses the issues of benefit sharing, risk and benefit balancing, trust and the motivation for people to donate samples. These issues are examined in the light of sustainable long-term policies in which trust plays a major role. Although in the light of their need to attract research funding, it is difficult for scientists not to raise high expectations. This, however, may damage the trust of the public in the biomedical establishment and hurt the research in the long run. Comparing biobanks in Japan and the UK, Masui makes clear that although Japan and the UK share problems, they solve them with different strategies, reflecting the diverging historical and institutional contexts of these countries.

Chapter 5, by Mary Miu Yee Waye and Connie Ho, uses the example of biobanking for dyslexic studies in Hong Kong to help decision-makers think about the complex decision of whether to invest in biobanking in Hong Kong. The authors identify various issues of trust and privacy related to the biobanking of genetic and health data on dyslexia in Hong Kong. These include the awareness of the benefits of storing genetic data of dyslexic children and worries about privacy. Some parents are hesitant about finding the cause of their child's learning problem and worry about the labelling effect of genetic data. The authors treat this labelling effect as a cultural issue of genetic discrimination, possibly leading to the family 'losing face' in the community. On the other hand, without a national biobank, a complex multifactorial disorder such as dyslexia is not likely to be solved soon. For the current existence of multiple laboratories, each collecting its own population of dyslexic subjects and associated data on cognitive profiles, has led to a duplication of efforts and forms an inefficient way of using research resources. Nevertheless, without more research findings on the pattern of heritability and interaction between genotypes and phenotypes, the authors are reluctant to recommend linking a genetic database with a healthcare database or with the education systems in Hong Kong.

Part II pertains to biobanking in developing Asian countries. In Chapter 6, Margaret Sleeboom-Faulkner analyses the e-discourse surrounding the dream of a Dutch scientist, Dr X, to create an HGDP in Bandung, Indonesia. Not only did this scientist hope to fulfil his dream of placing Indonesia on the genetic world map, he also aspired to set up the largest biomedical research centre in Indonesia, using blood samples from other laboratories, medical centres and the jungle. The most important programme would study genes for familial syndromes present in Indonesia (such as cancer and mental retardation), some of which may actually be specific for ethnic groups. Much has been written on the targets of the HGDP, but the incentives for and motives of geneticists to set up biobanks are little understood. Many works have ideologically 'framed' the biobanking initiatives of scientists as those of imperialists, Orientalists, racists and positivists or hailed their initiatives as those of explorers of humanity, pioneers of science and saviours of mankind. By analysing his e-letters to fellow scientists, academics and politicians, this chapter discusses the shortcomings of theories that variously ascribe to Dr X a colonial attitude, imperialist powermongering, scientific aspirations, biomedical ambition and an entrepreneurial spirit. This chapter, instead, proposes to analyse the factors of socio-cultural identity, political strategy, institutional convention and rhetoric as the elements of an institutional approach to understand the targets of biobanking. Such an approach should incorporate an assessment of socio-economic inequalities, on both international and national levels, public healthcare needs, research regulation, and differences in academic culture from a comparative perspective.

Nandini Kumar, the Deputy Director General of the Division of Basic Medical Sciences of the Indian Council of Medical Research (ICMR), in

Chapter 7 discusses regulatory issues in relation to genetic sampling in India. In India the analysis of DNA samples has to be carried out by Indian scientists or laboratories and no sample can be sent out of the country without following official guidelines. Her great worry is that violations occur despite the measures taken to protect the human rights of the participating individuals and populations. Kumar in particular emphasises the regulatory efforts of the Department of Biotechnology (DBT) and the ICMR, which are combined in a programme to study human diversity and, in the near future, to undertake the establishment of a national repository. Even though the guidelines for obtaining samples regarding consent, access to samples, ownership rights of DNA, and quality control should suffice in the regulation of data collection, there are worries about data protection and informed consent procedures, especially when applied among 'India's 4635 castes, 3000 communities and 28,000 endogamous groups', among which genetic information could be a source of tribal conflict.

Kumar is especially concerned about the illegal export of DNA samples and the foreign exploitation of India's populations. She emphasises the important role of the government of Indian in regulating biomedical research in protection against the threat of foreign researchers exploiting rich Indian genetic resources. This view is complemented in Chapter 8, by Prasanna Patra and Margaret Sleeboom-Faulkner, on the Indian genomic biobank initiative from a community-based perspective, which focuses on sampling practices by Indian researchers 'at home' among tribal communities. Chapter 8 discusses India's establishment of a national biobank or National Repository (NR) from ethical, legal and social points of view. The aim of setting up the NR was to provide a centralised repository for the unsystematically collected and stored bio-specimens and health data at various research centres, including those in universities, public and private hospitals, and research institutes.

Chapter 8 starts with a description of how biological specimens and health data are gathered, generally, through two processes: first, through genetic epidemiological and carrier-screening programmes in communities and, second, from individuals and patients in hospitals. At a community level, it is the population geneticist, local hospitals and physical anthropologists that have been involved in data generation; at a hospital level, it is individual patients who provide data through undergoing testing or tissue donations. The investments into India's proposed national biobank initiative is expected to come under great scrutiny, owing to the enormous social, economic and healthcare disparity and a lack of well-defined and stringent regulatory mechanisms for the conduct of biomedical research. The aim here is to stimulate discussion on common bioethical issues, such as informed consent, confidentiality, benefit sharing and public trust, by referring to the views and practices of researchers and community leaders themselves. The bioethical issues discussed here are considered to be vital in the field of genomics in India from a community perspective based on an anthropological study.

The national issue is central in Chapter 9, by Wen-Ching Sung, who draws upon one year of field research in China. Sung argues that the genome projects in China mark the emergence of a new mode of nation-making, 'the bio-nation'. In the building of a bio-nation, the Chinese state views and mobilises the flora and fauna species as valuable genetic resources within its territory. Thus, nature is appropriated as pools of genetic data. Yet what has turned into bio-data is not limited to nature only. With the promise and hyperbole of the human genome projects, China's fifty-six ethnicities have become genetic depositories that may shed light on treating human diseases as well as tracing the evolution of human beings. In this context, biobanking in China is much more than a means to retrieve scientific data for academic research. In fact, it has been advocated as an effort towards preserving national resources. The issues of distrust surrounding the discourse and practice of biobanking in China thus connote different meanings from that in North America and Europe. While the discussions on distrusting biobanking in the West concern the confidentiality and protection of individual donors, the counterpart discourses in China stress the need to keep the biobanks within its border in order to prevent foreign companies and academics from stealing the nation's genetic resources. This chapter contextualises the development of biobanking in China.

Complementing Sung's chapter on biobanking policies in China, Xinqing Zhang's Chapter 10 focuses on the attitudes and perceptions of Chinese genetic scientists regarding ethical issues in the process of creating genetic databases. Plans are being put in place for the systematic collection and use of human genetic samples in China. Chinese health authorities, however, have not set up clear legal frameworks or ethical guidelines for genetic research involving large numbers of human genetic samples. For this reason, the author felt that an estimation of the attitudes of genetic researchers towards bioethical questions related to genetic sampling could lead to useful insights. Zhang organised a nationwide mail survey with relation to human genetic databases in various Chinese cities, such as Shanghai and Beijing, with the purpose of identifying whether Chinese research communities were concerned with fundamental ethical topics, such as informed consent, privacy, ownership, commercial conflict of interests, and ethical review mechanisms. On the basis of 300 questionnaires filled in by Chinese scientists and Ethical Review Committee members, the survey shows that the creation of genetic databases has been surrounded by controversy in the Chinese biomedical community. The author feels that proper measures should be taken to improve the quality of the informed consent process and quality of review procedures in China.

Contextualising biobanking issues

Regarding the biobanking activities in the developing countries discussed in this volume, Mainland China, India and Indonesia, authors have indicated problems related to the application of bioethical principles, linked to the attitude of researchers themselves, the availability and workings of bio-ethics committees, and poverty and poor access to healthcare. Lack of stringent regulatory mechanisms for the conduct of biomedical research may be intrinsic to developing countries; not so much for a lack of awareness or education, but as a result of an incapacitating institutional environment. This, in turn, is not just due to poverty or a lack of financial and intellectual resources, but also to a government priority for policing bioresources on a national level, rather than providing the 'home' bases with a stable regulatory environment. As a result, be it through international collaboration such as joint ventures and cooperation between universities or through local biobanking projects, regulation for genetic sampling on a local level remains sterile.

This blatant lack of nationwide discussion on biobanking differs greatly from the nationwide discourses that have been orchestrated in wealthy Asian nations with a high level of healthcare and welfare facilities. While discourses of distrust of the West in China and India stress the need to keep the bioresources within their borders, the discussions on confidentiality, trust, benefit sharing, and the protection of privacy and autonomy have been main concerns. Naturally, issues related to trust, privacy and informed consent are of great importance in large developing countries as well, but issues of national consideration have superseded them. No doubt the presumption that foreign companies would take advantage of the differences in regulatory circumstances (Ch. 9) plays an important role, but one surmises that local companies may be capable of similar forms of exploitation of vulnerable communities. It is almost as if regulators and researchers, belonging to the national elite, are more concerned with the protection of national heritage and their disciplinary area of research than with the protection of individuals.

Countries with poverty and weak healthcare facilities have a larger proportion of the population prepared to cooperate in projects. Richer countries need to lobby with the public to get them to trust biobanks and regulators, especially when it involves the coupling of healthcare data with those of the biobank. As the cases of Taiwan and Japan show, this takes the form of finding ways of benefit sharing and persuading the people to trust the government (Chs 2–4). Poor developing countries with a weak implementation of legislation take top-down decisions and leave their unprotected population open to exploitation by their own educated elite when it benefits the nation as a whole (Chs 9 and 10). A poor developing country, such as Indonesia, with weak authority and infrastructures, is vulnerable to the biobanking needs of advanced biomedical research of both foreign powers and enterprises at home.

Wealthy democracies pay much attention to the issue of linking healthcare data to biobanks (Hong Kong, Japan, Taiwan). They are at pains to make biobanks acceptable to the public by discussing at length issues of information technology, data protection and benefit sharing. They depict the positive prospects of medical advancement, and are sensitive to issues of unemployment, insurance and discrimination. Even though none of the issues around biobanking can be called 'cultural' in themselves, the diverse circumstances in the more affluent Asian societies are called 'culturally different' to the extent in which they differ from the model of Biobank UK (Ch. 4). When comparing

Biobank UK with other biobanking projects in Asia, such 'cultural' difference seems to be a trifle compared to the dilemmas found in large Asian developing countries. In the case of developing countries such as China and India the building of national biobanks will require more than technology and investors alone. For in a situation in which basic healthcare facilities and registers are unavailable, propaganda and education campaigns may not convince people that biobanks are for their benefit. It is for these circumstances that we need to think of the financial and economic politics of bioethics, whether national or universal.

Notes

1 China, in October 2003, launched its first biobank ('China launches first biobank' *People's Daily*, 30 October 2003).
2 In Europe such collections include various sizes of private and public collections of samples of DNA, blood, tissues, cell lines, plasma and cellular samples (Hirtzlin *et al.* 2003).
3 http://www.bioethics.ntnu.no/biobanks/index.php?id=biobanks
4 http://www.fhi.no/eway/default0.asp?e=0&pid=225
5 http://www.decode.com
6 http://www.geenivaranu.ee
7 http://www.meb.ki.se/biobank/index_en.html
8 http://www.ukbiobank.ac.uk
9 http://http://www.p3gconsortium.org/genomeutwin.cfm
10 http://www.stn.org.sg/index.htm
11 http://www.p3gconsortium.org/Cartagene.cfm
12 http://www.alspac.bristol.ac.uk/welcome/index.shtml
13 http://www.westlakes.org/live/html/comcon/projhi.htm#nccgp
14 'China launches first biobank', *People's Daily*, 30 October 2003.
15 Xiao Chunjie, cited in *People's Daily*, 28 November 2000.
16 Sites in the DNA sequence where individuals differ at a single DNA base are called single nucleotide polymorphisms (SNPs). A haplotype is a pattern of SNPs on a block of inherited sets of nearby SNPs on the same chromosome.
17 In November 2004, the HUGO Pacific Pan-Asian SNP Initiative was established by scientists from China, India, Indonesia, Japan, Koreas, Malaysia, Nepal, Philippines, Singapore, Thailand and Taiwan, aiming to look at 50,000 SNPs in each study participant (see the homepage of the Genome Institute of Singapore, available at: www.gis.a-star.edu.sg/homepage/gismediapress.jsp?pid=33)
18 http//:www.genomenewsnetwork.org
19 http://www.decode.com
20 http://www.p3gconsortium.org
21 http://www.fhi.no/artikler/?id=47033
22 http://www.rediff.com/money/2005/jun/13gene.htm
23 Other genetic and archaeological evidence shows that there never was an evolutionary separation between races, and the presumption that a purer race existed in the past is incorrect (cf. Templeton 2003).
24 For a long list of declarations against the HGDP, see Macer 1997: note 6.
25 It should be pointed out here that race is a social category presumed to refer to 'nature' (cf. Wade 2002). For a discussion of natural, cultural and scientific categories in an East Asian context, see Lee 2003; Sleeboom 2004.
26 The definitions of the 'main races' vary slightly per author. For instance, Wilson *et al.* distinguish four races, including the African, Caucasian, Asian and Pacific

Islander races (Wilson *et al.* 2001); Risch *et al.* (2002) make a distinction between five main races, including African, Caucasian (Europe and Middle East), Asian, Pacific Islander (for example, Australian, New Guinean and Melanesian) and Native American.

27 DNA Analysis Reveals Taiwanese Have Ancestors on Mainland, *Beijing Times*, 14 December 2001.

28 According to M'charek (2000), a marker is an identifiable physical location on a chromosome whose inheritance can be monitored.

29 Another concept of population depends on the type of genetic markers and on their number. The more markers we use, the higher the chances that the difference between groups will become blurred, and the better we can distinguish between individuals.

References

Berg, K. (2004) DNA sampling and banking in clinical genetics and genetic research. In: Glasner, P. (Ed.) *Reconfiguring Nature. Issues and Debates in the New Genetics.* Aldershot: Ashgate, 137–40.

Bhardwaj, M. (2007) 'Biobanks, association studies and validity: ethical, legal and social challenges in Asia', *JIBL*, 4(1)2007: 133–45.

Burton, B. (2002) Opposition stalls genetic profiling plan for Tonga Human Genetics Alert, 18 February 2002, Inter Press Service, http://www.hgalert.org

Cavalli-Sforza, L.L. (2000) *Genes, Peoples and Languages.* London: Penguin Books.

Cavalli-Sforza, L.L., Wilson, A.C., Cantor, C.R. *et al.* (1991) 'Call for a worldwide survey of human genetic diversity: a vanishing opportunity for the Human Genome Project', *Genomics*, 11: 490–1.

Chu, Jiayou *et al.* (1998) 'Genetic relationship of populations in China', *J Proc Natl Acad Sci USA*, September 29; 95(20): 11763–8.

Clarke, A. and Parsons, E. (1997) *Culture, Kinship and Genes. Towards Cross-Cultural Genetics.* London: MacMillan Press Ltd.

Collins, F.S., Brooks, L.D. and Chakravarti, A. (1998) 'A DNA polymorphism discovery resource for research on human genetic variation', *Genome Research*, 8: 1229–31.

Enserink, M. (1998) 'Opponents criticize Iceland's database', *Science*, 282 (5390): 859.

Glazier, A.M., Nadeau, J.H. and Aitman, T.J. (2002) 'Finding genes that underlie complex traits', *Science*, 298: 2345–9.

Hedgecoe, A.M. (2005) *The Politics of Personalised Medicine: Pharmacogenetics in the Clinic: Cambridge Studies in Society and the Life Sciences.* Cambridge: Cambridge University Press.

Hirtzlin, I., Dubreuil, C., Préaubert, N. *et al.* (2003) 'An empirical survey on biobanking of human genetic material and data in six EU countries', *European Journal of Human Genetics*, 11: 475–88.

Hopkins, M. *et al.* (2006) 'Putting pharmacogenetics into practice', *Nature Biotechnology*, 24(4): 403, 410.

House of Lords (2001) *Science and Technology* (Fourth Report). UK Parliament.

Human Genome Organisation (HUGO) Committee (1995) Human Genome Diversity (HGD) Project: Summary Document, Report of the International Planning Workshop of the HGD Project, held in Porto Conto, Sardinia (Italy) on 9–12 September 1993. London: HUGO Europe.

ICMR (Indian Council of Medical Research New Delhi) (2000) *Ethical Guidelines for Biomedical Research on Human Subjects*, 34-8, http://icmr.nic.in

International HapMap Consortium (2003) The International HapMap Project, *Nature*, 426: 789–96.

Jayaraman K.S. (1997) 'India drafts law to protect bio-sources', *Nature*, 390: 108.

Knoppers, M. (2003) *Populations and Genetics: Legal and Socio-Ethical Perspectives.* Leiden: Martinus Nijhoff Publishers.

Lander, E.S. *et al.* (2001) 'Initial sequencing of the human genome', *Nature*, 409: 860–921.

Lee, S. (2003) 'Racial profiling of DNA samples: will it affect scientific knowledge about human genetic variation? In: Knoppers, M. (Ed.) *Populations and Genetics: Legal and Socio-Ethical Perspectives.* Leiden: Martinus Nijhoff Publishers, 231–44.

Lock, M. (1994) 'Interrogating the human diversity genome project', *Social Science & Medicine*, 39: 603–6.

Lock, M., Cunningham-Burley, S. Franklin, S. *et al.* (2005) 'Eclipse of the gene and the return of divination', *Current Anthropology*, 46: S47–S70.

Macer, D.R. (1997) Bioethics and genetic diversity from the perspective of UNESCO and non-governmental organization. In: Knoppers, B.M., Laberge, C.M., Hirtle, M. (Eds) *Human DNA: Law and Policy. International and Comparative Perspectives.* The Hague: Kluwer Law International.

McKie, R. (2000) 'The gene map of Britain, and how it could save your life', *The Observer*, 13 February: 16–17.

Majumder, P.P. (2000) Genes, diversities and peoples of India. In: Macer, D. (Ed.) *Ethical Challenges as We approach the End of the Human Genome Project.* Tsukuba: Eubios Ethics Institute, 20–33.

Mak J.C., Leung H.C., Ho S.P. *et al.* (2006) 'A new haplotype of PDCD1 is associated with rheumatoid arthritis in Hong Kong Chinese', *J Allergy Clin Immunol*, 117: 92–6.

Marshall, E. (1997) 'Gene prospecting in remote populations', *Science*, 278(5338): 565.

Martin, P. and Kaye, J. (2006) The use of large biological sample collections in genetics research: issues for public policy. In: Glasner, P. (Ed.) *Reconfiguring Nature. Issues and Debates in the New Genetics.* Aldershot: Ashgate.

Martin, P. *et al.* (2006) *False Positive? Prospects for the Clinical and Commercial Development of Pharmacogenetics.* Nottingham: University of Nottingham/ University of York.

M'charek, A. (2000) Technologies of Similarities and Differences. On the interdependence of nature and technology in the Human Genome Diversity Project, Leiden: Dissertation Leiden University. The commercial version (2005) *Human Genome Diversity Project: An ethnography of scientific practice.* Cambridge: Cambridge University Press.

National Human Genome Research Institute (2002) International Consortium Launches Genetic Variation Mapping Project, http://www.genome.gov/10005336

Nicholson, D. (2000) Banking on genes, *The Scientist Daily*, 4 December. http://www. biomedcentral.com/news/20001204/03/

Peterson, A.L. (2001) *Being Human. Ethics, Environment, and Our Place in the World.* Berkeley, CA: University of California Press.

Pilnick, A. (2002) *Genetics and Society: An Introduction.* Buckingham: Open University Press.

Pomfret, J. and Nelson, D. (2000), An Isolated Region's Genetic Mother Lode. Chinese

Human Genome Project Millennium Pharmaceuticals Harvard School of Public Health, *Washington Post*, Wednesday, 20 December 2000, A01.

Radio Australia, 18 November 2000, *Europe Pacific Solidarity Bulletin*, 8 (4/5), *PacNews*, 28 November 2000. Available at http://www.antenna.nl/ecsiep/bulletin/bull8-4/6.html

Risch, N., Burchard, E. Ziv, E. *et al.* (2002) Categorization of humans in biomedical research: genes, race and disease, *Genome Biology*, 3, 1 July . The electronic version of this article is the complete one and can be found online at: http://genomebiology.com/2002/3/7/comment/2007

Rose, H. (2001) 'The commodification of bioinformation', a report prepared for the Wellcome Trust, www.wellcome.ac.uk/

Sándor, J. (2003) *Society and Genetic Information. Codes and Laws in the Genetic Era*. Budapest: Central European University Press, CPS Books.

Senituli, L. and Boyes, M. (2002) 'Whose DNA? Tonga & Iceland, biotech, ownership and consent', Australasian Bioethics Association Annual Conference, Adelaide, February 14–16 2002, http://www.wcc-coe.org/wcc/what/jpc/dna.html

Sleeboom, M. (2004) *Academic Nations in China and Japan: Framed in Concepts of Nature, Culture and the Universal*. London: RoutledgeCurzon.

Sleeboom, M. (2005) 'The Harvard case of Xu Xiping: exploitation of the people, scientific advance or genetic theft?' *New Genetics and Society*, 20, April: 57–78.

Sutrop, M. (Ed.) (2004) 'Trames', *Journal of the Humanities and Social Sciences*, 8: (1/2) Special issue: 'Human genetic databases: ethical, legal and social issues'.

Templeton, A. (2003) Human races in the context of recent human evolution. A molecular genetic perspective. In: Goodman, A.H., Heath, D. and Lindee, M.S. (Eds) *Genetic Nature/Culture. Anthropology and Science Beyond the Two-Culture Divide*. Berkeley, CA: University of California Press, 234–57.

Venter, J.C., Adams, M.D., Myers, E.W. *et al.* (2001) 'The sequence of the human genome', *Science*, 291: 1304–51..

Verma I.C. (1997) Ethical concerns in genome diversity studies in developing countries. In: Knoppers B.M., Laberge C.M., Hirtle M. (Eds) *Human DNA: Law and Policy International and Comparative Perspectives*. The Hague: Kluwer Law International, 257–63.

Verma, I.C. (2000) Ethics and human genome studies and genetic services in India. In: Fujiki, N. and Macer, D. (Eds.) *Bioethics in Asia*. Tsukuba: Eubios Ethics Institute.

Wade, P. (2002) *Race, Nature and Culture. An Anthropological Perspective*. London: Pluto Press.

Waldby, C. and Mitchell, R. (2006) *Tissue Economies. Blood, Organs, and Cell Lines in Late Capitalism*. Durham, NC: Duke University Press.

WHO (2002) Genomics and World Health. Report, http://whqlibdoc.who.int/hq/2002/a74580.pdf

Williams, G. and Schroeder, D. (2004) Human genetic banking and the limits of informed consent. In: Glasner, P. (Ed.) *Reconfiguring Nature. Issues and Debates in the New Genetics*. Aldershot: Ashgate, 148.

Wilson J.F., Weale M.E., Smith A.C. *et al.* (2001) 'Population genetic structure of variable drug response', *Nat Genet*, 29: 265–9.

Zhao T. and Tsung Dao-Lee (1989) 'Gm and Km allotypes in 74 Chinese populations: a hypothesis of the origin of the Chinese nation', *Human Genetics*, 83: 107.

Part I

Biobanking in welfare societies

2 Public trust, commercialisation, and benefit sharing

Towards a trustworthy biobank in Taiwan

Hung-En Liu and Terence Hua Tai

Some features of biobank research

Since the completion of the Human Genome Project, functional genomics has become a major focus in genetic studies. To understand the functions of genes and probe the complex interplay between genetic and environmental factors in causing common diseases, many researchers believe that large-scale biobanks are especially useful. Collections of biological specimens, medical records, and genealogical data become very valuable for biomedical research, especially in the areas of pharmacogenomics and population genetics (Kaiser 2002). Many countries, including Iceland, the UK, and Estonia, have launched ambitious biobank projects that are backed by government agencies and/or public funding (Austin *et al.* 2003). Taiwan has also done so.

In May 2005, Taiwan's government decided to sponsor a pilot project conducted by the Institute of Biomedical Sciences, Academia Sinica, Taipei. The pilot project is expected to lead, after three years of feasibility studies, to the official development of a national-level biomedical database that will in the long run store blood samples, genetic information derived from them, and personal lifestyle information, with linkage to health data and medical records, collected from 200,000 people aged 40–70 years. Potential participants will be randomly selected on the basis of household records and approached for informed consent (Lee 2004). The database is designed for research on the genetic and environmental factors in the aetiology of common diseases in Taiwan. According to the pilot project's proposal, already approved by Taiwan's National Science Council, there will be three recruitment centres located respectively in Miao-li county (in the central part of Taiwan), Chia-yi city (in the south), and Hua-lien county (in the east). The three locations were selected on the grounds that they together reflect the ethnic diversity of Taiwan, and that the healthcare infrastructure in these places is adequate enough to facilitate the informed-consent process, the collection of biological samples, and the acquisition of health information about participants in the future. Different combinations of ethnic groups are targeted: in Miao-li county, people of Fukien [Fujian] (nineteenth century or earlier), Hakka (nineteenth century or earlier), and Mainland (twentieth century) descent; in Chia-Yi [Jiayi] city, people of Fukien [Fujian] and Mainland descent; and in Hua-lien county, people from indigenous, Fukien, and Mainland descent. Though already approved, the pilot project is required by the National Science Council *not* to get off the ground unless a proper ethical and legal regulatory framework is put in place, and for good reasons.

Compared with traditional biomedical research, biobank projects of this magnitude have some unprecedented features. To begin with, they aim to collect tissue samples and personal genetic data from a very large population. For instance, while Taiwan Biobank plans to collect blood samples and health data from 200,000 people, the UK Biobank forecasts a cohort of at least 500,000 men and women aged 45–69 years from the UK population (UK Biobank 2002). As Michael Yeo noted, the more extensive the collection, the more the collection becomes an issue not only for the individuals but also for the population as a whole, and the greater the challenge may be for recruitment and data security (Yeo 2004).

Second, these biobank projects need a great deal of funding, logistic support, and technical collaboration, and therefore the public sector depends increasingly on private industry to fund and participate in the research. For instance, the Iceland government relies on a private company (deCODE) to establish a large-scale biobank (Greely 2000; Liu 2004), the Estonian Genome Project Foundation set up EGeen Inc. to market products of the Estonian biobank to the global pharmaceutical industry (Austin *et al.* 2003), and the UK Biobank also explicitly claims that 'involvement of the pharmaceutical and biotechnology industry in the project is essential' (UK Biobank 2002). In Taiwan, the biobank project involves private industry from an early stage. Many pharmaceutical and IT companies, including HP, IBM, and Vita Genomics, have eagerly pushed the government to establish a Taiwan Biobank (Ho 2004; Kuo 2004), and the government also clearly claimed that 'one main purpose of the biobank project is to promote the biotech and IT industries in Taiwan' (Chao 2004; Chou H-Y 2005; Hsieh 2004).

Third, while commercial involvement seems inevitable, it is expected that biobank projects will benefit society as a whole in terms of improving healthcare services and medical knowledge. Participant recruitment and sample collections usually appeal to the goodwill and altruism of individual donors. Considering the large number of participants needed, and given that large-scale biobanks are usually backed by public funding and/or government agencies, there is a general consensus that the success of the biobank project depends crucially on public trust and support (Hansson 2005; Petersen 2005; Sheremeta 2003).

Altruism of the public versus commercial involvement: the tension between them and some misgivings

Large-scale biobank projects and population genetics have triggered a number of controversies at both national and international levels. Much attention has been focused on issues of informed consent, privacy, and data security. This chapter suggests that we should pay more attention to some issues related to the above-mentioned features of biobanks, especially the issues that derive from the tension that easily arises between commercial involvement and the appeal to the altruism of the public.

Many people seem to have misgivings about commercial involvement in biomedical research, and various empirical studies show that these misgivings may affect public trust and support. For instance, public opinion data from Canada suggest that the public generally lacks trust in corporate responsibility in the biotechnology field and that it tends to mistrust researchers if they are collaborating with for-profit companies (Sheremeta 2003). In the UK, a survey by the Human Genetics Commission (HGC) in 2000 found a clear aversion to the use of personal genetic information for commercial purposes. The same survey also shows that British people favoured the public ownership of new products developed from using genetic information (Human Genetics Commission 2000). In Taiwan, according to a survey conducted by G-T Chou in 2005, 77.52 per cent of the interviewees worried about the possibility that their genetic information might be released for commercial purposes (Chou G-T 2005).

It is understandable that donors may feel betrayed or even cheated if they find that researchers or private companies appeal to altruism to collect their samples and/or data, on the one hand, but make profit and *do not actually return a reasonable portion of the profit to the public*, on the other. This was at stake in a recent lawsuit against a researcher and Miami Children's Hospital (MCH) filed by the Canavan Foundation and families afflicted with Canavan disease. This case involves an alliance between parents and not-for-profit organisations who sought the help of researchers to develop prenatal and carrier testing for Canavan disease, and they hoped that these diagnostic tools could be made accessible and affordable to the public. From the beginning, it was obvious that they donated their blood samples and money for the common good. That is exactly why they felt betrayed and cheated when they found out that, unbeknownst to them, the researcher and his employer, MCH, had obtained a patent for the Canavan disease gene that they had discovered. Subsequently, they had begun to charge royalties, thereby limiting the availability of the diagnostic tools in question. If the researcher and MCH had returned the benefit to the public and not applied for a patent, the plaintiffs might not have filed a lawsuit (Kolata 2000).

In addition to its possible adverse effects on public trust, commercial involvement may harm scientific integrity too. For instance, biomedical researchers have a tradition of free inquiry and free exchange of ideas, and objectivity is central to the scientific pursuit of truth. However, many empirical studies show that this tradition has been eroding because of the trend towards commercialisation (Krimsky 2003, Sheremeta 2003). Studies find that when a researcher has a financial interest in or funding by a company, results of his or her research tend to favour the sponsor's product, and are less likely to be published (because the sponsor may prohibit the researcher from publishing the findings that may have to be kept confidential in the interest of the sponsor), or at least are more likely to be published at a delayed time (because the sponsor may want to apply for a patent first) (Bekelman *et al.* 2003, DeAngelis 2000). Withholding data and findings from colleagues becomes more common, and many researchers are required by their sponsors (pharmaceutical companies) to do so because of commercial secrets or competition (Campbell 2002). Commentators also worry that the focus of biomedical research will be skewed away from basic research to what is potentially very profitable (Sheremeta 2003; Williams and Schroeder 2004). In the end, the decline of scientific integrity may further hurt the public trust in biomedical research and researchers.

Still other concerns are discussed in the sections below. As collaboration between industry and academia has been increasing and even encouraged by the government, conflict of interest situations may emerge more frequently. The public sector's increasing dependence on private enterprise to fund and participate in biobank research provides new opportunities for conflicts of interest to arise (Sheremeta 2003). Might the research participants be harmed because of researchers' conflict of interest? Is it fair or equitable to ask the public to donate for the common good, on the one hand, and yet let the researchers/companies have all the profits they make, on the other?

Researchers' conflict of interest

A conflict of interest is a situation where financial and other personal considerations have the potential to compromise or bias professional judgement and objectivity. The 'interest' does not necessarily mean 'financial interest'; it can be reputation, promotion, or even the interest of advancing science. A conflict of interest can happen at both individual and institutional levels. While the investigators/researchers may have their own conflicts of interest in human subjects research, there can be institutional conflicts of interest too if the interests of an institution or any of its influential officials may affect, or reasonably appear to affect, institutional processes, including the conduct, review, or oversight of human subjects research (American Association of Medical Colleges 2003; Association of American Universities 2001).

The American Association of Medical Colleges (AAMC) has made two reports to provide guidelines and recommendations for oversight of conflicts of interest in human subjects research (AAMC 2003). As the reports point out:

> [Though] competing interests … are an inescapable fact of academic life, … financial interests in human subjects research are distinct from other

interests inherent in academic life ..., because financial interests are discretionary, and because the perception is widespread that they may entail special risks. Specifically, opportunities to profit from research may affect – or appear to affect – a researcher's judgments about which subjects to enroll, the clinical care provided to subjects, even the proper use of subjects' confidential health information. Financial interests also threaten scientific integrity when they foster real or apparent biases in study design, data collection and analysis, adverse event reporting, or the presentation and publication of research findings.

While the guidelines mentioned above and government controls of conflict of interest in the USA focus only on financial interests (AAMC 2003; Henderson and Smith 2002; Steinbrook 2004), there are growing concerns about non-financial (intangible) conflict of interest issues, especially the issues of physicians' involvement in biomedical research (see, e.g., Caulfield and Griener 2002). Nowadays, biomedical research projects rely increasingly on physicians for participant recruitment and sample collection, and many physicians recruit their own patients to participate in their research (Caulfield and Griener 2002; Anderlik 2003). This research practice may be problematic because when a physician recruits his own patients for research, there can be inherent conflict of interest due to his dual roles: a physician's primary responsibility is to ensure the welfare of his patients, but as a researcher he expects to finish the research fast and smoothly to advance science, gain reputation, or even make a profit by collaborating with a company. When a physician recruits his own patients for research, the ethical issues include the potential for misleading the patients or even for coercion. Questions arise as to whether the patients can tell the difference between treatment and research. In fact, even if the patients can tell the difference, in many cases they may not dare to say no to the physician's invitation to participate in the research because they tend to be afraid that refusal may offend the physician or affect the treatment they will receive. When recruitment for research takes place at the same time the patient is asked to give consent to surgery, the substantial stress may exert undue influence on the patient and could be coercive (Anderlik 2003).

In May 2005, a 'scandal', as it was called in the *Apple Daily* newspaper, happened in a very prestigious military hospital in Taiwan. Several patients' families claimed that a physician collected blood samples from the patients for research without obtaining consent beforehand. However, the physician retorted that the patients had actually signed informed consent forms in advance (Lin 2005). This is a typical case of the confusion of patients and the possible conflict of interest arising from the physician's dual roles, a case which we can examine from at least three perspectives. To begin with, can the mere signatures on the forms represent genuine 'informed consent'? This is highly questionable because the patients may have confused the consent to participation in research with the consent to treatment, not to mention the possibility that they may have consented under the kind of undue influences mentioned above. Second, even if the physician has obtained the patients' informed consent, the information or knowledge gap between the physician and the patients makes it dubious that informed consent is sufficient to protect the patients. Shouldn't there be a mechanism that can reliably monitor the physician's conflict of interest and ensure autonomy and safety of the patients? Finally, did the physician notice the likelihood of conflict of interest due to his dual roles? If physicians are generally insensitive to this issue, shouldn't there be a training programme in research ethics for physicians and other biomedical researchers?

Unfortunately, in the past, some researchers in Taiwan intentionally made use of the patients' confusion mentioned above in an effort to collect blood samples or conduct research. For example, a physician in a well-known hospital in southern Taiwan ran a clinical trial on his patients which the patients thought was part of their therapy: the patients alleged that they had never known it was a clinical trial. In fact, the physician never denied their allegation, and he was later forced to resign from that hospital (Chen 2003). Another problematic research practice is that some researchers collected blood samples by offering at the same time a 'free health check' with the intention to induce lay people, especially the Taiwanese aborigines, to participate in medical research. Since the aboriginal communities in Taiwan usually lack adequate healthcare infrastructure and resources, this so-called 'free health check' did attract many inhabitants. However, some of these researchers only used the free health check as a mere 'bait' for promotional purposes, and never returned any health check results to the aborigines. According to a news report, 'one aborigine had been taken blood samples in the name of free health check from eight different research teams, but none of them ever told him the results', and 'thereafter, these aboriginal communities strongly distrust any healthcare professionals and medical researchers' (Chang 2001). The incident might actually involve a project that was explicitly intended to collect blood samples for research while at the same time promising to provide a health check for free. But not only did the aborigine mentioned in the news report seem to complain that he never received the health check results, but also, worse still, that he was unaware that the health check was offered by people intending to collect blood samples.

Even the research practices of some very prestigious institutions should have paid more careful attention to ethical concerns about this unsettling mixture of sample collection with free health checks. For instance, the 'Super Control' study – a small-scale biobank project – carried out by Academia Sinica in 2002–03 leaves room for the suspicion that it has unnecessarily put emphasis on a 'free health check' in a promotional letter sent to potential participants before project personnel went to their houses to ask for their consent to participation. This letter was the only information the potential participants would receive about the project before the coordinators' visits, but it only briefly mentioned the purposes and nature of the research after highlighting, in the very first paragraph of the letter, the provision of a 'free health check' (Institute of Biomedical Sciences, Academia Sinica 2002). Intentionally or not, the information provided in the letter *could* be misleading and unduly influence lay-people's perceptions of, and their decision on, participation in the research. In addition, the practice of visiting potential participants' houses directly and asking for their consent to participation right at the first visit is also questionable, because it is rather intrusive and may not give them enough opportunity to consider the content, nature, and risk of the research. Academia Sinica is in charge of the pilot project for Taiwan Biobank; for this reason, its research practices and relevant ethical issues should be carefully re-examined before inauguration of the project.

Moreover, the pilot project of Taiwan Biobank will collaborate with local hospitals and community physicians in collecting blood samples and medical records. The project should take great care not to put the physicians in a conflict of interest situation such as the one mentioned above. It must be noted that not only does physicians' conflict of interest tend to violate patients' autonomy, but it also tends to cause harm to patients because the physicians' professional judgement on the well-being of their patients may be compromised by their other roles or interests.[1] Some even worry that clinical care may be manipulated to meet the needs of the biobank (Anderlik 2003).

In the USA, many regulations and court decisions have been made to control these conflict of interest situations. For instance, in a 1990 landmark case, Moore v. Regents of the University of California, the Supreme Court of California stated that:

> we hold that a physician who is seeking a patient's consent for a medical procedure must, in order to satisfy his fiduciary duty and to obtain the patient's informed consent, disclose personal interests unrelated to the patient's health, whether research or economic, that may affect his medical judgment (793 P.2d 479, Cal. 1990).

By contrast, Taiwanese physicians, biomedical researchers, and policy makers seem to be unaware that the conflict of interest issue could be very important from the perspectives of public trust and participants' autonomy and safety. In the past ten years, there has been only one article whose title contained the term 'conflict of interest' and which discussed this issue in a biomedical context (Tsai 2004). Many physicians and dentists are insensitive to this issue, and a few of them even publicly recommended healthcare products and toothpaste in TV advertisements on behalf of some pharmaceutical companies. As collaboration between industry and academia has been increasing, and the distinction between private and public sectors is blurred in the area of biomedical research, conflict of interest situations will be seen more frequently (Henderson and Smith 2002, Kolata 2000). Thus, in the future, the government and/or professional groups in Taiwan should enact regulations or guidelines to require researchers to disclose their conflict of interest to research participants and the institutional review boards (IRBs); in fact, this requirement is already provided in articles 13 and 22 of the World Medical Association (WMA) Declaration of Helsinki (WMA 2000). Other monitoring mechanisms and ethical training programmes for researchers may be necessary as well.

The inevitable commercial involvement versus the necessity of benefit sharing

Though commercialisation of biomedical research may have an adverse impact on public trust and the protection of participants' autonomy and safety, commercial involvement in biobanking is inevitable and even necessary. First of all, one of the main objectives of population genetic research is to develop new drugs and treatments for human diseases. The pharmaceutical industry will inevitably be involved in the process, and it will play a crucial role in the translation of results from basic research into tangible products and procedures that may benefit individuals and society (Sheremeta 2003). Moreover, as discussed earlier in this chapter, because of the large scale of biobank projects, they need a great deal of funding, logistic support, and technical collaboration. In every country that plans to establish a population-based biobank, the public sector depends increasingly on private industry to fund and participate in the research.

While commercial involvement in biobanking is inevitable and even necessary, we have argued in this chapter that the conflict of interest situations should be curbed and mechanisms be designed to monitor the researchers' conduct and to protect the participants' safety. Moreover, in order to ensure the autonomy of potential participants and public trust, participants and the public should be made aware that possibilities for commercial exploitation exist in addition to any benefits for all, such as improvements in healthcare. If researchers always appeal to the 'common good' to lead the public's attention away from possible (or even inevitable) commercial involvement, then public trust and support will surely decline in the long run.

Nevertheless, merely requiring disclosure of conflict of interest, adherence to proper procedures for informed consent, and establishment of monitoring mechanisms may not be enough to gain donors' trust and increase public support of biobank research. These measures seem to be basically negative: 'it can only stop unethical or unwanted research from being undertaken, but cannot pro-actively steer the usage of a DNA bank' (Williams and Schroeder 2004: 98).

In our view, attention should also be paid to positive measures for sharing benefits with the participating populations.[2] For one thing, since biobanks rely on both the altruism of donors and government support for their establishment, and the public invests a significant degree of trust in the researchers and/or their commercial partners, it is necessary to design a mechanism to vindicate this trust and let the operation of biobanks match the donors' altruism by managing and using the biobanks at least in part for the common good. As long as the researchers solicit sample donations from the public by appealing to altruistic motivation, they have a responsibility to ensure that biobanks will be used for publicly endorsed ends (Williams and Schroeder 2004: 97). Benefit-sharing arrangements made before collection of samples may also ensure public trust and support because potential participants can actually know what the 'common good' will be.

There is another important reason for sharing benefits with the sampled populations. In the context of large-scale biobanking, important interests of various communities may reasonably be held to be at stake. Although personal identifiers of samples collected and stored in a biobank will be encrypted, subsequent research using the samples and genetic information derived from them will often depend on the availability of group identities – such as ethnic, gender, and occupational identities – of sample sources. Although such research holds promise for enormous improvements in medical knowledge and healthcare services, it also raises serious concern that publicised research results and their implications about the genetic and environmental factors in the aetiology of diseases might foster stigmatisation and unjust discrimination against vulnerable communities. Since an entire population may have to assume such risks, it is equitable that there should also be benefits for the population as a whole. A benefit-sharing mechanism balances the commercial interests with interests (and burdens) of sampled populations in a way that both pays due respect to, and reflects fairly, the relative contribution of whole populations to the research endeavours (Deschênes and Cardinal 2003; Sheremeta 2003).

Some biomedical researchers and companies may argue that their commercial success will automatically lead to the 'common good', because improved knowledge, new drugs, and new treatments developed from biobank data already count as benefits to the entire society. However, we would rebut this argument by noting that vulnerable participating communities, such as aboriginal or poor ones, will never actually benefit from the pharmaceutical companies' new drugs or treatments as long as they cannot afford them, as they very likely cannot if no appropriate measures are taken for sharing the fruits of genetic research using biobank resources. What these communities lack may be a basic healthcare infrastructure and fundamental medical services. It is highly doubtful that new drugs or improved knowledge can benefit them and improve their conditions without social arrangements that meet the demands of distributive justice.

Recently, benefit sharing in genetic research, especially in biobank research, has gradually become an emerging legal requirement in international law (Deschênes and Cardinal 2003; Sheremeta 2003). Many international documents, such as the HUGO Ethics Committee 'Statement on Benefit Sharing' (2000), the WHO report 'Genetic Databases: Assessing the Benefits and the Impact on Human and Patient Rights' (2003), and the UNESCO 'International Declaration on Human Genetic Data' (2003), have strongly called for benefit sharing with participating populations in genetic studies. Nevertheless, benefit sharing can take different forms and is subject to varying societal and cultural values. It is not necessarily monetary. What constitutes 'benefit' and 'sharing' would depend on needs, values, priorities, and cultural expectations. The public should be consulted on the issue of benefit sharing before collection of samples gets under way.

Some biobank projects have already made arrangements for benefit sharing. For instance, in Iceland, deCODE and Roche reached an agreement that should Roche develop any products as a result of the database research, it would provide these products free of charge to Icelanders during the period of patent protection (Greely 2000). In Canada, Newfound Genomics promised to return a percentage of net profits to an independent foundation set up by the company for the population (Knoppers 2000).

It is noteworthy that article 19 of the UNESCO International Declaration on Human Genetic Data (2003) provides the following:

> [B]enefits resulting from the use of human genetic data, human proteomic data or biological samples collected for medical and scientific research should be shared with the society as a whole and the international community… . [B]enefits may take any of the following forms: (i) special assistance to the persons and groups that have taken part in the research; (ii) access to medical care; (iii) provision of new diagnostics, facilities for new treatments or drugs stemming from the research; (iv) support for health services; (v) capacity-building facilities for research purposes; (vi) development and strengthening of the capacity of developing countries to collect and process human genetic data, taking into consideration their specific problems; (vii) any other form consistent with the principles set out in this Declaration.

Such provision requires not only that benefits from genetic research using population biobanks be shared within a society, but also that they be shared internationally. Although this requirement of global justice may sound too high-minded, it is one that developing countries should not fail to press, on grounds of reciprocity or even of the human genome as 'common heritage', if they involve themselves in collaboration with other countries in the collection and utilisation of human genetic samples.

We want to stress at this point the importance of grounds of *reciprocity* for benefit sharing. Large-scale biobanking creates a cooperative scheme, which involves not only research institutions and profit-seeking companies, but also the population(s) from which individual sample sources come. Even if recruitment appeals to altruistic motivation on the part of donors so that they should expect no material gains in return for themselves *as individuals*, demands for benefit sharing with the sampled population(s) as a whole are ethically justified because every *cooperative* scheme yields benefits and burdens that must be distributed fairly or equitably, on grounds of reciprocity, among stakeholders involved in the scheme. This notion of a cooperative scheme as created by large-scale biobanking and sustained through reciprocity points further to the need of public consultation, in addition to individual consent, on any large-scale biobank project through properly designed, democratic procedures of public deliberation (Tai 2005).

What proper arrangements for benefit sharing should be designed for Taiwan Biobank is as yet left an open and admittedly difficult question. However, from public debates and newspaper commentaries on Academia Sinica's pilot project for Taiwan Biobank, it seems clear that a consensus has already been established concerning the need for benefit sharing, and this is already a big step forward. Public consultations on the project are presently expected to get under way. While the aforementioned UNESCO International Declaration on Human Genetic Data suggests some ways to bring substance to the idea of benefit sharing, people in Taiwan must work out for themselves through democratic, deliberative procedure whether they can place trust in the biobanking project and, if so, how Taiwanese society in general, and the least fortunate social groups in particular, can eventually benefit from it, at least in terms of better and affordable healthcare services.

Conclusion

Most attention to biobanks from the ethical and legal perspectives has been focused on issues of informed consent, privacy, and data security. This chapter argues that we should pay greater attention to issues related to the trend of commercialisation of biomedical research and to the increasingly important but difficult problems raised by conflict of interest. In addition, the tension that seems hard to avoid between commercial involvement and the appeal to the altruism of the general public may, if not adequately dealt with, adversely affect public trust and support. We believe that benefit sharing with the populations involved is essential if such a tension is to be resolved or lessened in an acceptable way, and public trust and support is to be ensured.

In the future, the administrators or custodians of Taiwan Biobank should consider reaching an agreement with each company that applies for the usage of the data stored in the biobank. The agreement should include an article that provides that the company shall share benefits with Taiwanese society as a whole, to ensure that the Taiwanese people receive the 'common good' they have been promised.

Notes

1   At times, the consequence can be serious. For instance, in 1999 an American boy, Jesse Gelsinger, died because of the complications of an experimental gene therapy treatment administered by a physician who was a primary stakeholder in a bioge-netics company that would stand to profit from the experiment and new technology. In fact, this physician's employer, the University of Pennsylvania, also had a finan-cial interest in relation to the experimental study, so there was an institutional conflict of interest too. See Smith and Byers 2002.

2   It must be noted that this article argues the necessity of benefit sharing with the participant *population*, not the *individual participants*.

References

American Association of Medical Colleges (AAMC) (2003) 'Protecting subjects, preserving trust, promoting progress', *Academic Medicine*, 78: 225–45.

Anderlik, M.R. (2003) 'Commercial biobanks and genetic research: ethical and legal issues', *American Journal of Pharmacogenomics*, 3: 203–15.

Association of American Universities (AAU) (2001) 'Report on individual and institutional financial conflict of interest', http://www.aau.edu/research/COI.01.pdf

Austin, M.A. *et al.* (2003) 'Genebanks: a comparison of eight proposed international genetic databases', *Community Genetics*, 6: 37–45.

Bekelman, J.E. *et al.* (2003) 'Scope and impact of financial conflicts of interest in biomedical research', *Journal of American Medical Association*, 289: 454–65.

Campbell, E.G. (2002) 'Data withholding in academic genetics: evidence from a national survey', *Journal of American Medical Association*, 287: 473–80.

Caulfield, T. and Griener, G. (2002) 'Conflicts of interest in clinical research: addressing the issue of physician remuneration', *Journal of Law, Medicine and Ethics*, 30: 305–8.

Chang, L.-W. (2001) (in Chinese) 'Aborigines suffer because genetic decoding blooms', *China Times*, Taipei, 19 March.

Chao, P.-R. (2004) (in Chinese) 'Taiwan will establish a biobank', *Economic Daily News*, Taipei, 25 February.

Chen, H.-M. (2003) (in Chinese) 'Patients became guinea pigs', *China Times*, Taipei, 4 November.

Chou, G.-T. (2005) 'Risk governance of biobank', paper presented at the 2nd Annual Meetings of Taiwan Bioethics Association, Kaohsiung.

Chou, H.-Y. (2005) (in Chinese) 'Government will spend 15 billion to build Taiwan as an island of biomedical technology', *Economic Daily News*, Taipei, 7 April.

DeAngelis, C.D. (2000) 'Conflict of interest and the public trust', *Journal of American Medical Association*, 284: 2237.

Deschênes, M. and Cardinal, G. (2003) 'Survey of national approaches to the development of population genetic biobanks', http://cbac-cccb.ca

Greely, H.T. (2000) 'Iceland's plan for genomics research: facts and implications', *Jurimetrics Journal*, 40: 153–91.

Hansson, M.G. (2005) 'Building on relationships of trust in biobank research', *Journal of Medical Ethics*, 31: 415–18.

Henderson, J. and Smith, J. (2002) 'Financial conflict of interest in medical research: overview and analysis of federal and state controls', *Food and Drug Law Journal*, 57: 445.

Ho, W.-L. (2004) (in Chinese) 'Taiwan should establish a biobank as soon as possible', *Economic Daily News*, Taipei, 30 April.

Hsieh, P.-H. (2004) (in Chinese) 'Government will promote Taiwan Biobank', *Commercial Times*, Taipei, 25 February.

Human Genetics Commission (2000) 'Report to the human genetics commission on public attitudes to the uses of human genetic information', http://www.hgc.gov.uk/UploadDocs/DocPub/Document/public_attitudes.pdf

Institute of Biomedical Sciences (2002) (in Chinese) *Manual for Coordinators' Training in the Super Control Study*, Taipei: Academia Sinica.

Kaiser, J. (2002) 'Population databases boom, from Iceland to the US', *Science*, 298: 1158.

Knoppers, B.M. (2000) 'Population genetics and benefit sharing', *Community Genetics*, 3: 212.

Kolata, G. (2000) 'Sharing of profits is debated as the value of tissue rises', *New York Times*, New York, 15 May.

Krimsky, S. (2003) *Science in the Private Interest*, Lanham: Rowman & Littlefield.

Kuo, Y.-C. (2004) (in Chinese) 'Taiwan will build a Center of Chinese Genetic Data', *Liberty Times*, Taipei, 19 March.

Lee, T.-Y. (2004) (in Chinese) 'Taiwan has decided to establish a biobank', *China Times*, Taipei, 25 February.

Lin, Y.-M. (2005) (in Chinese) 'Doctor at Tri-Service G. Hospital collected blood of patients without consent', *Apple Daily*, Taipei, 22 May.

Liu, H.-E. (2004) (in Chinese) 'A study on the legal policy of Iceland's population databases and biobanks', *Taipei University Law Review*, 54: 45–99.

Moore v. Regents of the University of California, 793 P.2d 479 (Cal. 1990).

Petersen, A. (2005) 'Securing our genetic health: engendering trust in UK Biobank', *Sociology of Health and Illness*, 27: 271–92.

Sheremeta, L. (2003) 'Population biobanking in Canada: ethical, legal and social issues', http://cbac-cccb.ca.

Smith, L. and Byers, J. (2002) 'Gene therapy in the post-Gelsinger era', *JONA'S Healthcare Law, Ethics and Regulation*, 4: 104.

Steinbrook, R. (2004) 'Conflicts of interest at the NIH-resolving the problem', *New England Journal of Medicine*, 351: 955.

Tai, T.H. (2005) 'Informed consent and benefit sharing in the context of human biobanking', paper presented at the 2005 ELSI Symposium on the Legal Implications of Biobanking, Taipei.

Tsai, D.F.C. (2004) (in Chinese) 'Conflict of interest in medical healthcare', *Journal of Taipei Medical Association*, 48: 35.

UK Biobank (2002) 'Protocol for the UK Biobank', http://www.ukbiobank.ac.uk

Williams, G. and Schroeder, D. (2004) 'Human genetic banking: altruism, benefit and consent', *New Genetics and Society*, 23: 89.

World Medical Association Declaration of Helsinki (2000) Ethical Principles for Medical Research Involving Human Subjects.

Yeo, M. (2004) 'Biobank research: the conflict between privacy and access made explicit', http://cbac-cccb.ca

3 Biobanks in Japan
Ethics, guidelines and practice

Gerard Porter

Introduction

This chapter presents the regulatory framework for human biobanks in Japan as it has been set out in two recent sets of guidelines; the Fundamental Principles of Research on the Human Genome (issued in June 2000) and the Ethical Guidelines for Analytical Research on the Human Genome/Genes (issued on 29 March 2001 and subsequently revised in full on 28 December 2004 and revised in part on 29 June 2005). Where appropriate, these guidelines are compared and contrasted with the UK Biobank project's Ethics and Governance Framework, and points of convergence and significant differences between the Japanese and British approaches are highlighted.

An attempt has been made to discuss the main streams of thought in the Japanese academic literature and to draw out possible weaknesses or areas of uncertainty within the Japanese approach, as well as its strengths. Whilst every effort has been made to ensure the accuracy of Japanese-English translations, the author accepts full responsibility for any changes in nuance which may have occurred. The chapter provides an update to the author's 2004 paper 'The Regulation of Human Genetic Databases in Japan'.[1]

Background to the Japanese guidelines

The regulators who have drafted the Ethics and Governance Framework of the UK Biobank project[2] have attempted to achieve a complex and delicate balance of interests.[3] The regulatory structure that they have devised must maximise the usefulness of the Biobank as a long-term resource for a variety of population-based genetic studies. At the same time, the Ethics and Governance Framework must also strive to protect the rights and dignity of the donors of genetic material on whose highly sensitive information this research will be based. When assessing how the various competing interests have been balanced, it may be useful to observe the ways in which other jurisdictions have attempted to regulate human genetic databases – to analyse the successes and problems of their regulatory frameworks and see how behaviour has evolved in those countries in practice.[4] Viewed in this spirit, Japan may prove to be a valuable and informative case study in the regulation of human biobanks.

In both Japan and the UK, the state-sponsored development of the bioinformatics sector has increased the need for public trust and support for medical research; but in both Japan and the UK, this has come at a time when public faith in the medical and research communities is at a particularly low point following a series of high-profile medical scandals. Further factors have raised significant additional challenges for the drafters of the Japanese guidelines, giving them perhaps an even more arduous task than that of their British counterparts. Privacy concerns amongst the Japanese public are perhaps more acute than in the UK, due to hypersensitivity about potential genetic discrimination and the general aversion in Japanese society to discussing the issues raised by inheritable disease. 'Western' concepts of bioethics have been introduced into Japan only relatively recently, and whilst some medical professionals have actively sought to reduce paternalistic attitudes and to integrate more patient-centred decision-making into their clinical practice, leading Japanese bioethicists have nevertheless been critical of the current rather limited reception of bioethical principles amongst the Japanese medical establishment. The perception in the eyes of the Japanese public of a medical research community that is reluctant to change in spite of widespread pressure serves only to further entrench an atmosphere of mistrust. Some commentators had suggested that the combination of these factors would likely serve to discourage public participation in biobank projects in Japan.[5]

In order to demonstrate some credible mechanism for protecting the rights of research subjects and win back a degree of public trust, two sets of guidelines were introduced in Japan to regulate the collection, utilisation and storage of human genetic material: the Fundamental Principles of Research on the Human Genome (the Fundamental Principles) created by the Bioethics Committee of the Council for Science and Technology, which were published on 14 June 2000;[6] and the Ethical Guidelines for Analytical Research on the Human Genome/Genes (the Ethical Guidelines) issued on 29 March 2001 by the Japanese Ministry of Health, Labour and Welfare (MHLW), the Ministry of Education, Culture, Sports and Technology (MEXT) and the Ministry of Economy, Trade and Industry (METI),[7] which were revised in full on 28 December 2004, and revised in part on 29 June 2005.[8] Both sets of guidelines aim to achieve public understanding and to develop a secure and fair environment for genetic research.[9] The two major themes which run through the guidelines are *autonomy* and *privacy*; these have been articulated in the forms of *informed consent* and *anonymisation*, respectively.

One (admittedly crude) method of assessing the practical success of the guidelines would be to attempt to measure the extent to which they have contributed to an increase in participation rates in genetic research projects. In 2005, Dr Yusuke Nakamura, who leads the large-scale BioBank Japan project, reported that 87 per cent of patients approached and asked to donate biological samples for BioBank Japan had agreed to do so.[10] This surprisingly high figure demonstrates that, at least in the specific context of BioBank Japan, the earlier fears about Japanese genomic studies being hampered by low rates of public participation have not been borne out in practice.[11]

However, this is not to say that the guidelines themselves are perceived as being a resounding success on all fronts. Indeed, Japanese commentators have pointed to numerous problems with the current framework. First, it has been suggested that the guidelines grant too much power to individual Ethics Review Committees, and fail to set clear limits on the exercise of their discretion.[12] Second, the guidelines have also been criticised for not clearly delineating the circumstances under which samples may be linked to research subjects and correlated with medical records, environmental data and other personal information that could potentially be used to identify a particular individual, or the conditions under which such sensitive data may be transferred to external organisations or subcontractors.[13] Third, some observers have pointed to the fact that these guidelines are not hard law but *gyōsei rippō* (administrative legislation), a commonly used Japanese administrative technique that is often favoured in fast-moving areas such as biotechnology where formal legal regulation is frequently perceived to be too restrictive.[14] Fourth, the adequacy of the mechanisms currently in place for monitoring and enforcing the guidelines has also been questioned.[15] Although the MHLW does back its guidelines with the threat of sanctions such as the withdrawal of research funds from institutions that violate its provisions, it is unclear what normative impact the guidelines will have on private research institutes that are not dependent on state funding.

At a broader social level, the regulatory culture surrounding biomedical issues in Japan has also been criticised, as the current lack of public debate and discussion about advances in human genetics is seen by some as being symptomatic of an overly paternalistic culture towards decision-making and policy in the medical sphere. Some Japanese observers have suggested that there is a need for more public debate with regards to biobanks and their benefits in order to win public support and understanding.[16] The UK Biobank project has been cited in Japanese academia as being laudable in this respect.[17]

Government funding of biotechnology in Japan

The Japanese government has identified the growth and development of its biotechnology industry as being vital to ensuring Japan's continued economic prosperity in the twenty-first century. Biotechnology's privileged status has prompted a series of well-funded government initiatives, which aim to enhance the international competitiveness of Japan's biotech sector and to apply scientific advances to bring benefits to the health and welfare of the Japanese population.[18]

As part of the current drive, the Japanese government has funded the creation of a multitude of new SNP databanks and also aims to consolidate and incorporate a number of earlier bioinformatics and database projects (which were initiated in the pre-genome-sequencing era of the late 1980s and early 1990s) into larger databases.[19] Of the new databases currently being assembled, the BioBank Japan project is perhaps the best known, but a number of smaller studies established under the auspices of the Millennium Projects also promise to further knowledge significantly in the field of human genomics.[20]

BioBank Japan

The large-scale BioBank Japan project commenced in June 2003 with funding of US$180 million from MEXT.[21] The project aims to identify the genetic causes of forty-seven common diseases (including heart failure, asthma, cerebral infarction and various forms of cancer) and to pinpoint genes related to adverse drug reactions.[22] It is envisaged that the information yielded through this research will be utilised as the basis for the development of new drugs and diagnostic methods, and that this will eventually contribute to the establishment of a new era of 'personalized medicine'.[23]

The project will collect samples of blood and sera from 300,000 patients suffering from the targeted forty-seven common diseases. As of April 2007, samples had been collected from 178,724 patients at sixty-six participating hospitals located throughout Japan.[24] Analysis of candidate disease loci and genotyping of associated SNPs takes place at Riken's SNP Research Center in Yokohama. BioBank Japan is led by Dr Yusuke Nakamura,[25] who also serves as Director of the University of Tokyo's Human Genome Center[26] and was the principal investigator for Japan on the International HapMap Project.

Whilst the large-scale, state-sponsored nature of BioBank Japan suggests obvious parallels with the UK Biobank, three significant differences exist in terms of the scientific objectives and designs of the Japanese and British projects. First, whereas BioBank Japan confines its focus to the investigation of forty-seven specific diseases, it is expected that the UK Biobank will be used for the study of a far wider range of medical conditions.[27] Second, whilst BioBank Japan is designed as a 'retrospective study' and therefore analyses samples obtained from patients who already exhibit symptoms of specific conditions, the UK Biobank, is a 'prospective study' that will analyse the samples of 500,000 individuals aged between 40 and 69 and track their health over the long term.[28] As such, few participants will be suffering from serious, debilitating disease at the time of enrolment in UK Biobank. A final major difference between the two projects is the the UK Biobank's particular emphasis on elucidating the interaction between genetic, environmental and lifestyle factors in the aetiology of disease. To this end, the UK Biobank will supplement data derived from analysis of stored blood and urine samples with information from the participants' medical records and other health-related and lifestyle data.[29] By contrast, it appears that at the current time BioBank Japan researchers have not been given MEXT authorisation to obtain environmental or lifestyle information from project participants, even though some clinical data regarding participants is gathered and stored.[30]

Encouraging public participation in genetic research: the politics of trust

In Japan, as elsewhere, the success of BioBank Japan and subsequent projects depends not only on the quality of the science and technology to be applied in genetic analysis, but also upon the ability to secure *public participation* – an issue which in turn hinges upon engendering *public trust*. When viewed from this perspective, Japan's current push into bioinformatics development comes at something of an inopportune moment for the Japanese government, as trust in the medical profession is at an historic all-time low.[31] A number of high-profile medical scandals have severely undermined public faith in the ethical integrity of the once highly respected medical establishment, which is now generally perceived by the Japanese public as being 'unable to regulate itself'.[32]

A further important factor which may discourage biobank participation in Japan is the fear of genetic discrimination in the context of marriage, employment and insurance. This anxiety is highly accentuated in Japan, where hereditary disease has traditionally been stigmatised and even discussion of inherited conditions is generally avoided.[33] Incidents of medical institutions selling patients' medical information (complete with full medical history, address and telephone number) to pharmaceutical companies and pharmacies have not helped promote public trust with regards to the way the medical profession protects confidential information.[34] In 1999, the Japanese Medical Association issued a warning that this particular attitude towards genetic disease in Japan, combined with the current mistrust with which Japanese medical and research professionals are regarded, would be likely to result in a general reluctance to participate in genetic research programmes, particularly if a credible and secure system of personal data protection could not be demonstrated to the Japanese public.[35]

The Japanese regulatory framework for human biobanks

The Fundamental Principles and the Ethical Guidelines were introduced in Japan in order to bring much needed clarification of the ethical standards governing genetic and genomic research. Both sets of guidelines bear a degree of similarity to the broad principles outlined in the UNESCO Declaration on the Human Genome and Human Rights.[36]

Two major themes which run through both sets of guidelines are *autonomy* and *privacy*, which have been articulated in the forms of *informed consent* and *anonymisation*. The most salient points shared by the guidelines are:

1   The use of informed consent forms to mediate almost all aspects of the relationship between researchers and research subjects.
2   The requirement for consultation at all key stages between research subjects and individual researchers or research groups, the head of

research institutions and Ethics Review Committees. Research must proceed upon the authorisation of the Ethics Review Committee and the permission of the heads of the research institute. Ongoing progress reports regarding the research, as well as its results, must be reported to the Ethics Committee.

3   The decision as to whether or not to participate in a programme of research should not result in any difference in terms of treatment for an individual, and consent may be withdrawn with no prejudicial effects towards the subject during such time period as withdrawal is possible.

4   Measures for the anonymisation and protection of personal data must be implemented.

5   The donation of biological materials is voluntary and without financial reward. Furthermore, intellectual property rights and other economic benefits which may arise as a result of the research will not be attributed to the participant.

Although the guidelines set the parameters for further information collection, they are not retrospective. The guidelines are not intended to apply to clinical laboratory tests designed to obtain genetic information for direct therapeutic benefits to the subject or his/her blood relatives.[37] Nevertheless, it is stipulated that such tests and analyses of the human genome should be properly conducted with due respect for the aims of the present guidelines and with reference to other guidelines established by relevant organisations.[38]

The Ethical Guidelines were revised in full on 28 December 2004 following a review of the provisions by the three ministries and an online public consultation.[39] The 2004 amendments sought to clarify and/or expand upon several points within the original text pertaining to, *inter alia*, the responsibilities of the heads of research institutes where genomic or genetic research is being conducted, the modalities of protecting personal information, various aspects of the process of obtaining informed consent and the operation of Ethics Review Committees.[40] Some further minor amendments to the Ethical Guidelines were made on 29 June 2005.[41]

Understandings and consent

Recruitment

Participation in biobanks and genome research in Japan will be entirely voluntary, and the collection of blood and tissue samples, etc., for research purposes may only proceed once the participant has given his or her informed consent.[42] This notion of voluntary participation is reinforced by Principle 5 of the Fundamental Principles, which states that '[a]n individual who is requested to provide a research sample but does not consent to that request should not be disadvantaged as a result of his/her refusal'.[43]

Informed consent of participants in genetic research

The reception of the doctrine of informed consent in Japan

In order to appreciate the significance of the emphasis being placed upon informed consent, it may be useful to locate the emergence of the two sets of guidelines within their particular historical and cultural context. Despite Japan being a signatory to the 1964 Declaration of Helsinki and a number of other international instruments, the integration of the concept of respect for patient autonomy, with informed consent as its primary expression, has been a slow and gradual affair. There remains significant diversity with regards to the degree of importance attached to the doctrine amongst members of the medical community.[44] More traditional and conservative institutions have shown little enthusiasm for change from more paternalistic approaches to patient care.

Nevertheless, a number of rulings from the Supreme Court have gradually established informed consent as a legal matter within Japanese medical jurisprudence.[45] In a 1981 ruling, the Japanese Supreme Court stated that physicians have a legal duty to explain the nature of and the risks involved with the surgery that is to be carried out and can only proceed once consent has been obtained.[46] A further significant legal development occurred in a 2000 Supreme Court ruling, where nominal damages were awarded purely for mental suffering as a result of the breach of the duty to obtain informed consent (and with no claim with regards physical harm) in a case where a doctor performed a blood transfusion on a patient despite her autonomous decision not to undergo the transfusion due to her religious beliefs.[47] Yet despite these Supreme Court decisions, some observers assert that the doctrine of informed consent is still at an early stage of its reception in Japan. Kimura's criticisms would suggest that the acknowledgement of informed consent as a legal matter has not necessarily translated uniformly into behavioural change or normative consensus with regards to the degree of significance that should be attached to the doctrine in a clinical setting.[48] However, this situation may change in light of the aforementioned Supreme Court rulings, though perhaps more for defensive legal reasons than because of unanimous agreement amongst the Japanese medical community on the ethical importance of respecting patient autonomy.

Ambivalence with regards to informed consent has been even more striking in the conduct of *ex vivo* medical research, with a number of incidents demonstrating what appears to be the complete lack of any kinds of ethical controls or respect for individual autonomy when gathering blood and tissue samples. For example, in 2000, researchers at the National Cardiovascular Centre in Osaka, the University of Kyushu, Fukuoka, and Tohoku University, Sendai, all admitted that they had used thousands of blood samples for analysis without obtaining prior informed consent from donors.[49] Events such as these received considerable attention in the Japanese media, further compounded public mistrust and fuelled the perception that the Japanese medical and scientific community has little regard for patient/subject rights.[50] A particular fear amongst Japanese researchers was that a lack of support from the Japanese public would force Japanese scientists to gather their samples and conduct research in other countries where proper legal safeguards ensure willing public participation.[51] At least one leading Japanese geneticist had complained of difficulties in obtaining tissue samples for research due to the scarcity of willing donors.[52]

The two sets of guidelines therefore came as a move to construct a framework which would enable scientists to conduct effective research while simultaneously protecting the rights of research subjects. Both sets of guidelines give extensive details with regard to the types of consent needed before DNA samples may be obtained and genomic analysis can be undertaken.

Defining informed consent in the context of genomic research

The central importance of informed consent in the context of genomic research is enshrined in Principle 5 (Basic Conditions) of the Fundamental Principles, which states that:

> A research sample may be collected from an individual subject for research on the human genome only after the participant has first been given a sufficient explanation of the research, and has given, of his/her own free will, his/her informed consent ... The consent should be expressed in writing.[53]

The Ethical Guidelines define informed consent in the following way:

> The consent voluntarily given for provision and handling of samples etc. by research subjects who are requested to provide them after having been explained in detail beforehand by the chief researcher about the significance, objectives and methods of planned research, anticipated outcomes and inconveniences and other relevant information and having fully understood such details. The present Guidelines require that informed consent be obtained in writing.[54]

These definitions of informed consent can be deconstructed to reveal the following three component elements:

1 *A duty to explain:* When the researcher is obtaining informed consent, the 'objectives, methods and potential outcomes of the research, and also any disadvantages and losses that the participant might incur, should be explained in a clear and understandable way'.[55]
2 *A duty to ensure that the patient has understood the explanation:* In order for consent to be valid, it is essential that the participants understand the information provided to them about the proposed project and the research

that will be conducted using their samples. When obtaining informed consent, researchers are required to describe the project in terms that are clear and readily understandable to non-specialists. The explanation should be provided in a number of steps, with the researcher taking time to ensure that each point has been understood by the participant. Allowing participants the time to read through and reflect upon written material is also seen as being desirable.[56]

3   *A duty to obtain informed consent in writing:* The consent should be given in writing and a suitable record kept. However, if a participant has difficulty or is unable to consent in writing for a particular reason, such as difficulty in writing or motor functions, appropriate alternative methods, e.g. audio-visual recording, should be prepared. It is stressed that alternative methods should be chosen in place of written consent only if there is such an impediment for the participant, and not for the sake of convenience for the researchers.[57]

A detailed explanation of the issues that must be specifically included in the informed consent form can be found in the Ethical Guidelines' Bylaw Regarding the Contents of the Written Information for Informed Consent.[58]

The scope of consent and 'comprehensive consent'

A recurrent problem with the use of the doctrine of informed consent in biobank projects is that often at the time that consent is given, it is not (nor cannot be) totally clear precisely what the subject is consenting to. In Japan as elsewhere, this becomes an important issue when it is anticipated that the same biological samples may be utilised in a number of genome analysis studies or integrated within larger databases. As a general principle, when participants grant consent they are consenting to the researcher carrying out research specifically within the framework of the particular research project that has been explained and outlined to them. Nevertheless, due to the value of samples to researchers, the Fundamental Principles permit a significant softening of the strictness of this requirement by permitting researchers to obtain '*comprehensive consent*', whereby consent is granted not only to a specific and defined project, but also extends to other genome analysis or to other related medical research. Typically, this will mean that the subject will consent to the use of the samples in a specific project and also for 'studies aimed at other purposes'.[59]

In accordance with the general principle of obtaining informed consent outlined above, detailed information should be given about other genome analyses and related medical research that are anticipated at that time, regardless of whether any of them are actually undertaken later. Nevertheless, it is stressed that 'comprehensive consent should not be solicited simply for the sake of convenience or saving of labour'. The control mechanism at this stage is the Ethics Review Committee, which should not approve a research project requiring comprehensive consent unless the confidentiality of the personal information of the participant (including genetic information and identifying information) is guaranteed.[60]

Broad versus specific consent: discussions in the context of Biobank UK

The issue of the scope of informed consent was a point of contention during the formulation of the UK Biobank Ethics and Governance Framework. Opinion was divided on the question of whether the Biobank could legitimately adopt a pragmatic approach whereby *broad consent* (i.e. consent to participate in UK Biobank with all that that implies)[61] would be obtained, enabling the Biobank to then utilise the sample in a wide range of genetic studies, or whether under the strictest interpretation of the principle of autonomy underpinning the Helsinki Declaration it would be necessary to obtain *specific consent* from participants for each and every research project making use of their samples and data in order for the consent to be valid and for the Biobank project to be ethically acceptable.[62] Most commentators on the 2003 UK Biobank Ethics and Governance Framework agreed that as long as sufficient explanation was given at the time of obtaining consent and participants then freely agreed to participate in the UK Biobank and all that that entails, there would be no need to go back to the participants to obtain new consent for each and every proposed genetic study.[63] In addition to imposing significant logistical and financial burdens upon the UK Biobank and being unduly troublesome to participants who may not want further contact, repeated recontacting would not be necessary, as most genetic research undertaken using the Biobank would fall within the scope of the broad consent given by participants (even though this would not constitute full and informed consent in its strictest sense). On this point, the Nuffield Council on Bioethics' views on the ethical acceptability of *genuine consent*, even if that consent is *not necessarily fully informed*, are highly instructive:

> The ethically significant requirement is not that consent be complete, but that it be genuine. Ensuring that consent is genuine is mainly a matter of care in detecting and eliminating lack of consent ... If all reasonable care is exercised, adequate and genuine consent may be established, although it will necessarily fall short of fully informed consent.[64]

Nevertheless, in order for the consent to indeed be genuine in accordance with the Nuffield Council's definition, some commentators believe that procedural mechanisms should be established to deal with particularly sensitive kinds of research projects (e.g. behavioural genetic research)[65] or requests for access to Biobank data by particular kinds of researchers (e.g. the tobacco industry),[66] which may fall outside the scope of even broad consent. The hostile reaction in Iceland to exclusive access to the Icelandic Health Sector Database being granted to a private company illustrates that although many people may be willing to provide samples and allow access to their medical records for a public venture which aims at improving the health and welfare of the nation, these sentiments may not necessarily equate with a willingness to participate in research undertaken by commercial entities aimed at generating profits.[67] Under the UK Biobank's current regulatory regime, if and when such ethically sensitive situations were to arise, the Ethics and Governance Council would decide whether or not access to the database should be granted and, if so, whether the particular research proposal falls outside the scope of the broad consent granted and if participants need to be recontacted and new consent obtained.[68]

Although, in fact, a similar mechanism of obtaining broad consent with the safeguard of an Ethics Review Committee to identify research proposals that fall outside the scope of consent was also adopted in Japan, the issues of ethically sensitive research and the involvement of the private sector have not drawn much discussion in Japan. In contrast to the UK Biobank position, no specific mention of the fact that the resultant databases may be used by commercial entities needs to be made in the informed consent form. As with the situation in the UK, it remains open to speculation quite how Japanese Ethics Review Committees will evaluate requests for access to genetic and medical data for use in potentially sensitive studies or to what extent Japanese participants might oppose the use of their genetic samples and medical information in certain kinds of private sector research.

Recruitment and consent in practice: BioBank Japan

Potential BioBank Japan participants are chosen from amongst the inpatients and outpatients at collaborating hospitals who suffer from one or more of the forty-seven diseases to be studied under the project.[69] Introductory information regarding the project is provided in the form of BioBank Japan posters[70] that are displayed in waiting rooms and information films that are played in rotation on waiting room televisions.[71] The initial approach to the patient is made by doctors, who then hand over the informed consent process to a 'medical coordinator'. The medical coordinator guides the patient to an interview room, where details regarding the research project are explained to them using explanatory literature and a video.[72] Patients are free to raise any questions they have about the project and what their participation will entail. If the patient consents to participate in the project, a blood sample is then taken. The informed consent process usually takes around 40–50 minutes.

In its review of current practice, the BioBank Japan ELSI Committee found that whilst the majority of collaborating hospitals had implemented effective measures to maintain the particpant's privacy during the enrolment process, this was not universal. In a few hospitals, personal information was taken in consultation rooms where the participants sat at adjacent booths separated only by a simple partition such as a curtain.[73] It is not clear if the exchanges between participants and medical coordinators could be overheard by other participants, or if they were covered by the general noise within the room. The BioBank Japan ELSI Committee is acting to rectify this problem by further standardisation of procedures at collaborating institutions. Interestingly, concerns regarding 'oral privacy' were also highlighted within the UK Biobank's Ethics and Governance Council report regarding the pilot project that involved around 4000 participants between March and June 2006.[74]

Collection of data from health-relevant records

The UK Biobank will gather information from the participants' medical records and other health-related data and correlate this information with ongoing genetic analyses. This process is generally seen as a valuable aid in enabling researchers to 'complete the health picture' and to increase the richness and accuracy of their analysis. The intention to link participants' samples to their medical records in the UK will be stated at the time of obtaining consent. One perhaps surprising aspect of genomic research in the context of the BioBank Japan project is that whilst some clinical data regarding the participant is gathered and stored, at the current time BioBank Japan researchers have not been given permission from MEXT to obtain information regarding donors' lifestyle patterns or environmental factors for correlative research, even in an anonymised form.[75] This anomaly might be explained by the aforementioned fear of discrimination and the hypersensitivity to the stigma related to genetic disease in Japan.[76] However, Masui and Takada have argued that research on samples alone without correlation to medical records and environmental factors is likely to be of limited scientific value, and have called for further discussion and clarification of the precise conditions and procedures for correlative research in order to maximise the value of research whilst ensuring that a robust system for protecting privacy is in place.[77] It seems that resolution of this key issue will emerge from subsequent discussions between the Ethics Review Committees and the various ministries.[78]

Provision of health information to participants

It is likely that some participants in genetic research studies may want to know the results of the analysis of their samples and the implications that these are likely to have for their health. Some participants may even expect this feedback as part of the *quid pro quo* for participation. Principle 13 of the Fundamental Principles establishes the general rule that each individual participant has the right to know about his/her own genetic information resulting from the research.[79] This should be explained to the participant when obtaining consent.[80]

Some general exceptions to the rule that a Japanese participant has the right to know about his/her genetic information are elucidated in both the Fundamental Principles and the Ethical Guidelines. In the following cases, genetic information need not be disclosed to the participant:

1   In the case of large-scale research projects, where analytical research on the human genome/genes is conducted to reveal a relationship between a certain disease and a gene or the function of a certain gene by comparing the genetic information of a large number of people or genes, and the genetic information of a single participant is not sufficient by itself to confer accuracy or reliability on a diagnosis of the condition of health or other medical aspects of the participant and thus disclosure to the participant would not be of sufficient significance;[81]

2   if the sample has been anonymised during the course of the research and can no longer be identified with the donor;[82]

3   if disclosure of some or all of the research results could have an adverse effect on the life, health, property or other rights and interests of the participant or third parties.[83]

The general policy adopted by the UK Biobank is that of *not informing* participants of the results of ongoing research. This is due to the potential logistical difficulties involved, the possibility of harming the participant by giving feedback outside the clinical setting and a desire to avoid the higher legal 'duty of care' that would be imposed on providers of clinical diagnoses as opposed to researchers merely taking samples.[84] However, some observers in the UK have expressed discomfort with a possible scenario of researchers becoming aware of a participant having a predisposition to a medical condition for which a treatment is available, but then being prevented from disclosing this information to the participant and informing him or her of the risk because of the Biobank's information disclosure policy.[85] As something of a compromise, UK Biobank participants may be informed of any clinically relevant findings elicited during the initial consultations with the Biobank research nurse (such as abnormal blood pressure) and then be asked to contact their GP about those findings. This very limited feedback policy has been questioned on both moral and legal grounds.[86]

The right not to be informed

As a general principle, the participant has the right not to be informed of his/her genetic information resulting from genetic research.[87] Accordingly, research findings may not be made known to the participant against his/her will.[88] Researchers are under an obligation to explain to participants about their right not to be informed when obtaining consent. However, even if a participant expresses an autonomous, *ex ante* desire not to be informed, a provision in the Ethical Guidelines allows researchers the option to recontact participants and reconfirm whether or not they wish to be notified of the results of genetic analysis in situations where it would be in their best interests to receive that information.[89] Factors to be taken into account by the Ethics Review Committee before giving authorisation for the researcher to recontact the participant include:

1　The effect on the lives of the participant and his or her blood relatives;
2　the availability of effective therapeutic measures and the health condition of the participant;
3　the likelihood of the participant's blood relatives having the same disease or medical problem;
4　the stipulation of informed consent pertaining to disclosure of the research results.

The Ethical Guidelines stipulate that the chief researcher should first give a 'sufficient explanation' to the participant and then confirm whether or not he or she would wish any relevant information to be disclosed. If the participant consents, then the chief researcher may impart the adverse research results; but if consent is not given, then the information must not be disclosed. The practical weakness with this approach is that by the very act of recontacting the participants to ask if they would wish to be informed of a possible health risk, their putative 'right not to know' their own genetic status has already been compromised to a certain extent.[90]

The withdrawal of consent

In Japan, under the Fundamental Principles and the Ethical Guidelines, the rights of research subjects to withdraw consent for the use of their samples and resultant data hinge upon the issue of whether a particular sample can be linked to the individual who donated it.[91] In principle, consent can be withdrawn at any point without disadvantage to the participant if the sample can still be attributed to its donor. In such cases, the sample itself and the data derived exclusively from it should be destroyed.[92] However, researchers are not obliged to destroy samples and data when the materials have been anonymised so that they can no longer be linked to the donor, or if there is very little possibility that personal information could be elicited from them if they were not destroyed and it would require a great deal of time and effort to destroy the samples, or if the research results have already been made public.

Information regarding the right to withdraw consent, and its limitations, should be provided in advance of obtaining the donor's informed consent.[93] When the period of preservation of samples etc. as prescribed in the research protocol has elapsed, the chief researcher should destroy them in accordance with the conditions agreed on with the participants or their representatives. This provision, however, does not apply to cases where samples etc. are provided to a human cell, gene or tissue bank or are commercially available.[94] Furthermore, the research results need not be destroyed in instances where they have already been made public.[95]

The two sets of Japanese guidelines do not make any provision for individual participants to selectively opt out of individual projects involving a particular kind of research, nor are there any alternatives to complete withdrawal of consent to participate, such as the three options of 'no further contact', 'no further access' and 'no further use' established within the UK Biobank Ethics and Governance Framework.[96]

Withdrawal of consent by the relatives of a deceased research subject

The issue of whether the relatives of a deceased participant in a biobank can exercise control over their deceased relative's samples and medical records by withdrawing the original consent for participation was considered in a 2004 ruling by the Supreme Court of Iceland.[97] The court held that the daughter of a deceased research subject who participated in a biobank was entitled to prevent the transferral of her father's medical data into a genetic database. The Court's reasoning was founded upon the argument that as the plaintiff and her deceased father both shared genetic characteristics, the information in the database would allow inferences to be made about the plaintiff herself. It was therefore held that her father's medical records fell within the scope of personal information protected by the plaintiff's constitutionally protected right to privacy.[98]

With regards to this sensitive issue, the Japanese approach is to stress that researchers should respect the feelings and wishes of the participant's surviving family and his/her blood relatives.[99] Consequently, the appointed representative of the participant may withdraw his/her consent for the use of the participant's sample at any time without penalty by expressing the intention of doing so in writing.[100] The dead participant's representative should be a person who is considered to be able to represent the will that the deceased was supposed to have while alive.[101] When consent has been withdrawn, the chief researcher should in principle anonymise and destroy the samples, etc., provided by the participant and the research results concerning the participant.[102] However, as with the withdrawal of consent by participants themselves, a number of limitations apply to a representative's right to withdraw from a research project. Consent can not be withdrawn if research samples have been anonymised so that they cannot be linked to the participant, or if there is very little possibility that personal information could be elicited from them if they were not destroyed and it would require a great deal of time and effort to destroy the samples, or if the research results have already been made public.[103]

The representative of a deceased research subject will first have to be appointed by the chief researcher of a research institution before he/she can exercise this right. It is not entirely clear how institutions will prioritise competing interests if a conflict of views should emerge amongst relatives.

Privacy

The protection of personal medical data in Japan

As the analysis of genetic information can reveal an individual's current physical condition and that individual's predisposition for developing genetic disease, there is the fear that these highly sensitive data could be used to discriminate against individuals in the context of employment or insurance. This anxiety is highly accentuated in Japan, where inheritable disease has traditionally been stigmatised and where even public discussion of the issues raised by inheritable genetic disorders is generally shunned.[104] A number of observers have therefore predicted that the issue of privacy, above all others, is likely to be the one which will most significantly shape public perceptions of genetic research and analysis in Japan.[105]

Confidentiality within the guidelines

One of the key themes permeating both sets of guidelines is *anonymity*. The Ethical Guidelines and the Fundamental Principles both stress that at all points during the collection, storage, transfer and utilisation of human biological materials, the anonymity and privacy of the research subject must be protected. Principle 11 of the Fundamental Principles outlines in broad terms the importance of protecting the confidentiality of personal genetic information.[106] The explanatory notes to the Fundamental Principles also impose an obligation on research institutions to establish a control system for the safekeeping and protection of the personal information related to providing samples, including the separate storage of information which can link the sample and the donor.[107]

The strict confidentiality of personal information is also enshrined as one of the basic principles of the Ethical Guidelines. The duties of researchers and heads of research institutions in protecting subject privacy are described as being the protection of personal information, efficiently responding to complaints about standards of data governance, and the prevention of the disclosure of personal information (unless this disclosure is justified).[108] Principle 12 of the Fundamental Principles provides that if a breach of confidentiality occurs, the participant who has sustained damages from that disclosure is entitled to receive compensation or indemnity.[109] However, these legal remedies are not elucidated in any great detail within the Fundamental Principles themselves.

The anonymisation of research samples

As a general rule, all samples should be anonymised before analytical research on the human genome/genes is conducted, in order to protect and maintain the privacy of the participants.[110] The Ethical Guidelines define two types of anonymisation:

1 *Linkable anonymisation*, where samples may still be linked to subjects. Samples etc. are made unidentifiable, but a table that links them with codes or numbers assigned to research subjects is retained to allow identification, if necessary.
2 *Unlinkable anonymisation*, where samples cannot be linked to subjects. Samples, etc., are made irreversibly non-identifiable without retaining a table that links them to codes or numbers assigned to research subjects.

A two-stage process has been constructed for the anonymisation of donor information. The first stage of anonymisation occurs when subjects donate materials to a medical institution, and a second level of anonymisation occurs if this information is transferred for incorporation into a larger, central database. In the first stage, samples taken and used in research by the medical institution retain information that can be used to identify the subject, but the procedures for linkage and the information necessary are tightly controlled by the personal information managers. Stuart describes the procedures for anonymisation and holding samples separately from identifiable medical and personal information in the following way:

> Managers appointed to supervise the protection of donor information provide each sample with a unique bar code, containing donor informa- tion. Details of the coding are kept on a stand-alone workstation, protected by fingerprint operated mouse controls and a hard disc drive password entry system. Paperwork relating to the donors is kept in a high-security safe. The bar coded sample is then passed (along with the age of the donor) to the research team.[111]

The second stage of anonymisation occurs when research data is transferred into a central database:

> [t]he data, along with the age of the donor, are fed into an online computer connected to the central database. Linkable now only by age and under normal circumstances amongst an array of data for similarly aged donors, the data and donor should no longer be unequivocally linkable. Transfer to the central database further diminishes the probability of being able to link data to the sample donor.[112]

However, some observers argue that perfect anonymisation, resulting in there being no possibility whatsoever of linking the samples or information to the donor, is extremely difficult. Whereas this generally constitutes the starting point of discussion regarding anonymisation within the UK, the great differ- ence with the debate in Japan is that it flows from the assumption that perfect anonymisation is, in fact, possible. This, according to Masui and Takada, influences the quality of the discussions in the two countries.[113]

Exceptions to the principle of protection of subject anonymity

Although the importance of the anonymisation of research samples is given considerable emphasis and is one of the principal mechanisms through which confidentiality is maintained, it is nevertheless accepted by the MHLW that some latitude should be maintained to allow the linking of research data to the sample donor in certain circumstances (e.g. when results indicate a disease for which an effective method of treatment is available). Linkage under such conditions requires the approval of the head of the research institute and the authorisation of the Ethics Review Committee.[114]

A second exception to the principle of anonymity outlined above is that samples may be used without anonymisation if the participant or his/her representative has consented to the provision of samples or genetic information for research without anonymisation.[115] The intention to use samples without anonymisation must be stated within the research protocol and approved by both the Ethics Review Committee and the head of the research institute. Here, the issue of privacy is to an extent recast as an issue of autonomy.

Disclosure of genetic information to persons other than the participant

In many cases, an adverse test result for a genetic disorder will have serious implications not only for the individual concerned, but also for his/her blood relatives, who may share the same genetic condition. The Fundamental Principles state that at the time of obtaining informed consent, researchers should explain to participants about the implications of the result of a genetic test for family members, and allow the participant to make a judgement beforehand with regards to whether they would wish blood relatives to be informed of the results of the participant's genetic analysis or whether they would prefer this information to remain confidential. In the event of a conflict emerging between the right of the participant to keep the information private and the right of blood relatives to know about elevated health risks, the Japanese regulatory framework retains the flexibility to give priority to the latter.[116] Thus, even in cases where participants have stated that they do not wish other parties or blood relatives to be informed of the results deriving from analysis of their genetic information, if a participant's genetic information has been found to indicate that there is a genetic effect which is very likely to endanger the lives of the participant's blood relatives and there is an effective therapeutic measure to deal with the genetic effect, the Ethical Guidelines authorise Ethics Review Committees to permit the disclosure of genetic information to the participant's blood relatives.[117] The chief researcher should consult with the Ethics Review Committee on the decision of whether or not to reveal the information to blood relatives, the extent of the information to be disclosed and the method of disclosure before informing blood relatives. These provisions therefore seem to grant researchers the discretion to inform blood relatives of the results of genetic analysis, rather than imposing an obligation to inform in all cases.

Whilst the Japanese guidelines mostly presume that blood relatives would desire to know the results of the participant's genetic test if it would impact on their own health, some provision is also made for protecting the interests of blood relatives who may not wish to know this information. The Ethical Guidelines stipulate that the chief researcher should first give a 'sufficient explanation' to the participant's blood relatives and then confirm that they wish the relevant information to be disclosed to them before imparting the adverse research results.[118] This approach gives rise to a similar tension to that which occurs when weighing whether or not to recontact participants who have stated that they do not wish to know their genetic status (i.e. that by the very act of contacting blood relatives to ask if they would wish to be informed of a possible health risk, their 'right not to know' their own genetic status has already been somewhat undermined).[119]

Adequacy of the current guidelines and possible future developments

Despite the Fundamental Principles and the Ethical Guidelines being described by the Japan Pharmaceutical Association as 'the strictest in the world' with regards to the regulation of genetic research, the two sets of guidelines have attracted some criticisms within Japan, most notably from the highly influential Japanese Medical Association (JMA).[120] The JMA has publicly stated that the current regulations are '[u]nclear on who owns genetic data, and leave too much discretion to advisory committees'.[121]

Since the publication of the two sets of Japanese guidelines, the methodology used to collect medical information in one particular epidemiological study has drawn a particularly hostile response from the JMA. Criticisms stemmed from the fact that the work of collecting personal lifestyle data for a genetics-related cancer study of 6000 residents of Kumano-cho, a small town near Hiroshima, was contracted out to ordinary citizens who were under no legal obligation to protect the confidentiality of the information they were handling.[122] The relevant provisions of the first draft of the Ethical Guidelines stated that personal information managers and assistant personal information managers must be individuals who have a legal duty to maintain the confidentiality of medical information, and must establish a clearly defined chain of command with privacy ensured at each stage.[123] However, no mention was made of the requisite status of staff employed to assist in the collection of information in genetic studies, or whether or not they must also be individuals who are under a legal duty to maintain the confidentiality of medical information. Whilst this omission was rectified by the 2004 revision of the Ethical Guidelines,[124] it has been suggested that the two sets of guidelines contain several such areas of potential uncertainty, and further revisions to clarify specific issues are expected.

The adequacy of the methods for the enforcement of the guidelines has also been called into question in Japan. The Ethical Guidelines require that every medical institution that intends to conduct genomic research must establish an Ethics Review Committee to screen and monitor research projects. However, as of 2002, many medical institutions had yet to fulfil even this basic obligation.[125] There is little up-to-date information on how this situation has progressed. Some commentators have stated that formal legislation backed with legal sanctions is urgently needed to regulate the collection, storage and use of human biological materials and genetic information. Whilst the various ministries acknowledge that the legislative route could offer some advantages in terms of enforcement, this is seen as being offset by the loss of flexibility and responsiveness offered by regulation through guidelines.

Conclusion

The regulatory framework for the UK Biobank has been described as being 'at an early stage in its evolution', with many areas lacking in detail.[126] This observation could perhaps just as easily be applied to the current Japanese regulatory provisions, where it seems that decisions on a number of sensitive ethical questions will made by individual Ethics Review Committees on a case by case basis. Stuart observes that clarification of key issues of data governance, such as determining the scope of informed consent, the linking of samples to participants' medical records and the transfer of identifiable information to external organisations without anonymisation, is likely to emerge from further dialogue between individual Ethics Review Committees and the Japanese MHLW.[127]

Finally, it has been suggested that attitudes towards bioethics are currently in something of a transitional stage in Japan. Observers have pointed to the uneasy coexistence of American-influenced concepts of bioethics (with an emphasis on individual autonomy) and the more paternalistic, traditional Japanese concepts of ethics (*rinri*), which Masui and Takada summarise as being the understanding that 'people in positions of responsibility possess knowledge which is correct, and we should be grateful when receiving the benefits of their wisdom'.[128] Both the effectiveness of the current regime for the regulation of human biobanks in Japan and the continuing evolution of Japanese attitudes towards bioethics require close monitoring over the coming years.

Notes

1 *SCRIPT-ed* (2004) 1(3): 449–487. http://www.law.ed.ac.uk/ahrc/script-ed/issue3/japan.asp (accessed 30 July 2007).
2 The UK Biobank is a population genetic database that will collect biological samples and data from 500,000 people currently aged 40 to 69.
3 Wellcome Trust, Medical Research Council, Department of Health (July 2006) 'UK Biobank Ethics and Governance Framework' Version 2.0. http://www.ukbiobank.ac.uk/docs/EGF_Version2_July%2006%20most%20uptodate.pdf (accessed 30 July 2007).
4 For commentary on the various tensions that have arisen in the context of population-based genetic databases, see Rose, H. (2006) 'From Hype to Mothballs in Four Years: Troubles in the Development of Large-Scale DNA Biobanks in Europe', *Community Genetics* 9(3): 184–189. See also Salter, B. and Jones, M. (2005) 'Biobanks and Bioethics: The Politics of Legitimation', *Journal of European Public Policy* 12(4): 710–732.
5 'Data Security is Crucial for Japanese Science' (2002) *Nature* 41: 689.
6 Bioethics Committee of the Council for Science and Technology (14 June 2000) 'Fundamental Principles of Research on the Human Genome'. http://www.mext.go.jp/a_menu/shinkou/seimei/gensokue.pdf (accessed 30 July 2007).

7 Ministry of Health, Labour and Welfare (MHLW), Ministry of Education, Culture, Sports and Technology (MEXT), Ministry of Economy, Trade and Industry (METI) (2001) 'Ethical Guidelines for Analytical Research on the Human Genome/Genes'. http://www.eubios.info/EGHGR.htm (accessed 30 July 2007).

8 MWLW, MEXT and METI, 'Ethical Guidelines for Analytical Research on the Human Genome/Genes' (29 March 2001; revised in full on 28 December 2004 and revised in part 29 June 2005). http://www.mext.go.jp/a_menu/shinkou/seimei/genome/04122801/all.pdf (accessed 30 July 2007). This most up-to-date version of the Ethical Guidelines has not yet been translated into English.

9 For a more sceptical appraisal of the Japanese government's motivation for drafting the guidelines and the rushed drafting process, see Tsuchiya, T. (2003) 'Provisions for Review for Genetic Research in Japan', *Eubios Journal of Asian and International Bioethics* 13: 127–130.

10 'Scientists Create Japan's Largest BioBank for Genetic Studies of 47 Common Diseases' (2005) 1(5 Affymetrix Microarray Bulletin at 3. http://www.microarray bulletin.com/community/article.php?p=32 (accessed 30 July 2007). Cf. Watts, G. (2007) 'UK Biobank Gets 10% Response Rate as it Starts Recruiting Volunteers', *British Medical Journal* 332: 1052.

11 The extent to which this high participation rate is directly attributable to the two sets of guidelines is, of course, open to question.

12 'Japan Guidelines under Fire after Protest Halts Study', (2003) *Science* 301: 1039.

13 Masui, T. and Takada, Y. (2003) 'Ethical, Legal and Social Issues of Genomic Research: New Phase of Genome Research Desperately Requires Social Understanding and Safeguards on the use of Medical Records and other Personal Information', *Yakugaku Zasshi* (*Journal of the Pharmaceutical Society of Japan*) 123(3): 107–119 (in Japanese).

14 Slingsby, B.T., Nagao, N. and Akabayashi, A. (2004) 'Administrative Legislation in Japan: Guidelines on Scientific and Ethical Standards', *Cambridge Quarterly of Healthcare Ethics* 13: 245–253.

15 Tsuchiya supra note 9.

16 Shimizu, K. (11 June 2003) 'Despite the Stakes, Public Role in Bioethics Debate Falls Short', *The Japan Times*.

17 Masui, T. (2003) 'Eikoku baiobanku keikaku no imi suru mono (The Meaning of the UK Biobank Project)', *Juristo* 1247: 29–36 (in Japanese). For criticisms of the UK Biobank for not facilitating sufficient debate amongst the scientific community regarding the scientific validity of the project's design cf. Barbour, V. (2003) 'UK Biobank: A Project in Search of a Protocol?' *The Lancet* 361: 1734–1738.

18 For discussion of the establishment of the Japanese Biotechnology Strategy Council in 2002, a body which provides advice on ways of further developing the Japanese biotechnology sector, see S&T section, British Embassy, Tokyo 'Japan's Biotechnology Strategy Council: Publication of Final Report. http://www.uknow. or.jp/be_e/science/reports/Life%20Sciences/34022X.pdf (accessed 30 July 2007). See also Biotechnology Strategy Council 'Biotechnology Strategy Guidelines (Draft)' (6 December 2002). http://www.biojapan.org/industry/reports/Biostrategy. pdf (accessed 30 July 2007).

19 Stuart, C. (2000) 'Human Genetic Databases in Japan', UK Parliament Select Committee on Science and Technology, Memorandum by the Science and Technology Section of the British Embassy, Tokyo at para. 9.2. http://www.publications. parliament.uk/pa/ld199900/ldselect/ldsctech/115/115we10.htm (accessed 30 July 2007).

20 For details of the numerous biobank projects in operation and being established in Japan see Stuart, ibid. at sections 3–6.

21 'Japan Launches Controversial Biobank Project' (2003) *Nature Medicine* 9: 982.

Further information about the BioBank Japan project can be found at http://www. src.riken.go.jp/eng/src/project/person.html (accessed 30 July 2007).

22 Morton, N.E. (2006) 'Fifty Years of Genetic Epidemiology, with Special Reference to Japan', *Journal of Human Genetics* 51(4): 269–277.

23 BioBank Japan is part of the broader MEXT Personalized Medicine Project which commenced in 2003.

24 http://biobankjp.org/info/sintyoku.html (accessed 30 July 2007) (in Japanese).

25 'Profile: Yusuke Nakamura' (2004) *Nature Medicine* 10: 560.

26 http://www.hgc.ims.u-tokyo.ac.jp/english/about.html.

27 These will include (but not be restricted to) cancer, heart disease, stroke, diabetes, and dementia. See UK Biobank: Protocol for a Large-scale Prospective Epidemiological Resource (21 March 2007) at section 1.1.1. http://www.ukbiobank.ac.uk/docs/UKBProtocolfinal.pdf (accessed 30 July 2007).

28 See Morton, supra note 22 at 276, comparing the British and Japanese projects and arguing that BioBank Japan's approach is more likely to yield useful therapeutic results. See also 'Huge Biobank Project Launches Despite Critics' (2006) *Nature* 440: 263; 'Beset by Practical Hurdles, UK Biobank Moves at Sluggish Pace' (2005) *Nature Medicine* 11(7): 696.

29 Supra note 27 at section 1.3.1.

30 Supra note 10 at 3.

31 Supra note 5.

32 'Doctors Far from Malpractice Accountability' (22 March 1999) *The Japan Times*; 'Japan: Convictions in Blood Scandal', (24 February 2000) Associated Press.

33 Stuart, supra note 19 at para. 1.6.

34 Kurata, N. 'Iryō jōhō no denshika to puraibashi saikō ('Reconsidering the Digitalisation of Medical Records and Privacy')'. http://www.fine.bun.kyoto-u. ac.jp/tr4/kurata.pdf (accessed 30 July 2007) (in Japanese); See also 'Byōreki tsuki meibo hanbai' ('Sale of name lists and medical records') (30 November 1999) *Asahi Shinbun (Asahi Newspaper)* (in Japanese).

35 Supra note 5.

36 UNESCO Universal Declaration on the Human Genome and Human Rights (11 November 1997). http://unesdoc.unesco.org/images/0011/001102/110220e. pdf#page=47 (accessed 30 July 2007).

37 Fundamental Principles, supra note 6, Explanatory Notes, Introduction, 2. Also see the Ethical Guidelines (as amended), supra note 8, Chapter 1 (Basic Ideas), 2 (Scope of Application) (1).

38 Ethical Guidelines (as amended) ibid.

39 Ministry of Education, Culture, Sports, Science and Technology, Ministry of Health, Labour and Welfare, Ministry of Economy, Trade and Industry, 'Concerning the Revision of the Ethical Guidelines for Analytical Research on the Human Genome/Genes' (28 December 2004) (in Japanese). http://www.mext.go. jp/a_menu/shinkou/seimei/genome/05011901.htm (accessed 30 July 2007).

40 For a comparison of the text of the first version of the Ethical Guidelines and the revised 28 December 2004 version, see the following comparative table (in Japanese): http://www.mext.go.jp/a_menu/shinkou/seimei/genome/05032201.pdf (accessed 30 July 2007).

41 Ministry of Education, Culture, Sports, Science and Technology, Ministry of Health, Labour and Welfare, Ministry of Economy, Trade and Industry, 'Concerning Understanding the Establishment and Operation of Ethics Review Committees Based on the Ethical Guidelines for Analytical Research on the Human Genome/Genes' (29 June 2005) (in Japanese). http://www.lifescience.mext.go.jp/bioethics/pdf/hito_genom_tuuti_h170629.pdf (accessed 30 July 2007).

42 Fundamental Principles, supra note 6, Principle 5 (Basic Conditions) (1).

43 Also see Ethical Guidelines (as amended), supra note 8, Chapter 3 (Basic Attitude

to Participants), 10 (Informed Consent), (11-Bylaw). The Japanese guidelines thus take the same approach as the UK Biobank Ethics and Governance Framework, which also states that '[p]articipation in UK Biobank will be voluntary', Version 2.0, July 2006, supra note 3, section I.A.1.

44 Asai, A. (1996) 'Barriers to Informed Consent in Japan', *Eubios Journal of Asian and International Bioethics* 6: 91–93.

45 Tejima, Y. (2002) 'Recent Developments in the Informed Consent Law in Japan', *Kobe University Law Review* 36(1): 45–59. See also Tejima, Y. (2001) 'Iryō ni okeru dōi no zentei toshite no setsumei gimu ni ihan shita tame ni mitomerareta isharyōgaku no santei ni kan suru kōsatsu' (Considerations Regarding the Calculation of Damages Awarded for Violation of the Duty to Inform as a Precondition of Consent in Medical Treatment)', *Juristo* 1199: 18–24 (in Japanese).

46 Supreme Court of Japan, 2nd petty bench, 1981.6.19., 1011 Hanji 54, 447 Hanta 78. The Supreme Court held, however, that a doctor's legal duty to inform the patient or their guardian of the risk involved with an operation did not to apply in situations where there were 'elements of uncertainty' with regards to the patient's condition, the likelihood of success of the operation or the prognosis without treatment. See Tejima 'Recent Developments in the Informed Consent Law in Japan', ibid. at 47–48.

47 The Supreme Court of Japan, 3rd petty bench, 2000.2.29., 54 Minshu 582, 1710 Hanji 97, 1031 Hanta 158.

48 Kimura, R. (1997) 'Images of Health and the Traditional Medicine: Bioethical Perspectives', *Zen Nihon Shinkyū Gakkai* (*Journal of the Japan Society of Acupuncture and Moxibustion*), 47(3): 93–100. http://www.bioethics.jp/licht thesis97_9-j.html (in Japanese) (accessed 30 July 2007). For discussion of the appropriateness of strong notions of patient autonomy as an ideal within a Japanese cultural context see Akabayashi, A. and Slingsby, B.T. (2006) 'Informed Consent Revisited: Japan and the US', *American Journal of Bioethics* 6(1): 9–14. See also Asai supra note 44.

49 'Japan Set to Tighten Ethics Rules for Genetic Sampling' (2000) 403 Nature 819–820. See also 'Kokuritsu junkankibyo senta ga idenshi 5000 ninbun o mudan kaiseki' (The National Center for Cardiovascular Disease analyses the DNA of 5000 Patients without Consent) (3 February 2000) *Mainichi Newspaper*. See also Tsuchiya supra note 9 at 129.

50 Supra note 5.

51 Ibid.

52 'Japan Set to Tighten Ethics Rules for Genetic Sampling', supra note 49.

53 Fundamental Principles, supra note 6, Principle 5 (Basic Conditions).

54 Ethical Guidelines (as amended), supra note 8, Chapter 6 (Definitions), 16 (Definitions), 7 (Informed Consent).

55 Fundamental Principles, supra note 6, Explanatory Notes, Principle 5 (Basic Conditions).

56 Ibid.

57 Ibid.

58 Ethical Guidelines (as amended), supra note 8, Chapter 3 (Basic Attitude to Participants), 10 (Informed Consent) (11-Bylaw).

59 Fundamental Principles, supra note 6, Principle 8 (Comprehensive Consent, and Unlinked and Unidentified Research Samples) 1.(a), 'If a participant consents to provide a research sample for genome analysis in a particular research project and, at the same time, anticipates and consents to the use of the same sample in other genome analyses or related medical research, then the research sample may be used for the latter "studies aimed at other purposes"'.

60 Ibid.

61 UK Biobank Ethics and Governance Framework, Background Document, Prepared by the Interim Advisory Group on Ethics and Governance (10 October 2003) section 1.B.1. www.wellcome.ac.uk/assets/WTD003287.doc (accessed 30 July 2007).
62 UK Biobank Ethics Consultation Workshop (25 April 2002) at 5. http://www.ukbiobank.ac.uk/docs/ethics_work.pdf (accessed 30 July 2007).
63 UK Biobank Ethics and Governance Framework: Summary of Comments on Version 1.0 (May 2004). http://www.wellcome.ac.uk/assets/WTD003285.pdf (accessed 30 July 2007).
64 Nuffield Council on Bioethics, Human Tissue: Ethical and Legal Issues (Nuffield 1995) at paras. 6.20–6.21.
65 See generally Nuffield Council on Bioethics, Genetics and Human Behaviour: The Ethical Context (Nuffield 2002).
66 For example, see 'Tobacco Firm to Profit From Cancer Genes', (12 November 2001), *The Guardian*; describing Japan Tobacco's efforts to use genetic research to develop and sell vaccine and antibody-based products aimed at the prevention and/ or treatment of lung cancer.
67 Supra note 4.
68 UK Biobank Ethics and Governance Framework, Version 2.0 (July 2006) supra note 3, section I.B.5.
69 Morton, supra note 22.
70 http://biobankjp.org/public/poster.html (accessed 30 July 2007).
71 Heisei 17 nendo ELSI iinkai katsudō hōkokusho (Report on the Activities of the ELSI Committee for 2005) (May 2006) (in Japanese). http://biobankjp.org/plan/elsi_report17.pdf (accessed 30 July 2007) at 8.
72 For inpatients with difficulty walking, the informed consent process can be performed at the patient's bedside.
73 Supra note 71 at 14.
74 UK Biobank Ethics and Governance Council, Annual Review 2006 (2006). http://www.egcukbiobank.org.uk/assets/wtx037217.pdf (accessed 30 July 2007) at 8.
75 Supra note 10 at 3.
76 Stuart, supra note 19 at 7.11.
77 Masui and Takada, supra note 13 at 115.
78 Stuart, supra note 19 at 7.10.5–7.11.
79 Fundamental Principles, supra note 6, Principle 13 (Right to be informed). The same general stance is established by Chapter 3 (Basic Attitude to Participants), 11 (Disclosure of genetic information), (1) of the Ethical Guidelines (as amended), supra note 8.
80 Ethical Guidelines (as amended), supra note 8, Chapter 3, (Basic Attitude to Participants), 10 (Informed consent) (11-Bylaw). The circumstances under which genetic information will not be disclosed must also be stipulated within the consent form.
81 Ethical Guidelines (as amended), supra note 8, Chapter 3 (Basic Attitude to Participants), 11 (Disclosure of Genetic Information) (1-Bylaw).
82 This second point is likely to be highly relevant in Japan, where as a general rule samples will be routinely anonymised before analysis is undertaken. See Stuart, supra note 19 at para. 2.5.
83 Ethical Guidelines (as amended), supra note 8, Chapter 3 (Basic Attitude to Participants), 11 (Disclosure of Genetic Information) (1-Bylaw). The same bylaw provides that chief researchers should exercise appropriate caution and sensitivity when explaining to participants the reason for non-disclosure of their genetic information. However, no concrete advice is given on how to implement this requirement in practice.
84 UK Biobank Ethics Consultation Workshop (25 April 2002), supra note 62 at 8; UK Biobank Ethics and Governance Framework, Background Document, Prepared by

the Interim Advisory Group on Ethics and Governance (10 October 2003), supra note 61, section 1.B.3.

85 UK Biobank Ethics and Governance Framework: Summary of Comments on Version 1.0 (May 2004) supra note 63 at section 3.3.3.

86 For critical discussion on the compatibility of the UK Biobank policy on patient feedback with both the common law duty of care and Article 2 of the European Convention on Human Rights, see Johnston, C. and Kaye, J. (2004) 'Does the UK BioBank Have a Legal Obligation to Feedback Individual Findings to Partici-pants?', *Medical Law Review* 12: 239–267.

87 See generally Laurie, G. (1999) 'In Defence of Ignorance: Genetic Information and the Right Not to Know', *European Journal of Health Law* 6: 119–132.

88 Fundamental Principles, supra note 6, Principle 14 (Right Not to Be Informed).

89 Ethical Guidelines (as amended), supra note 8, Chapter 3 (Basic Attitude to Participants), 11 (Disclosure of Genetic Information) (2-Bylaw).

90 See generally Laurie, G., (2002) *Genetic Privacy: A Challenge to Medico-Legal Norms*, Cambridge University Press. See also Andorno, R. (2004) 'The Right Not to Know: An Autonomy Based Approach', *Journal of Medical Ethics* 30: 435–439 at 436.

91 Fundamental Principles, supra note 6, Principle 10, (Withdrawal of Informed Consent).

92 Ethical Guidelines (as amended), supra note 8, Chapter 3 (Basic Attitude to Participants), 10 (Informed Consent) (10). The Explanatory Notes to Principle 10 of the Fundamental Principles, supra note 6, (Withdrawal of Informed Consent) provide that '[w]hen the consent to use a sample is withdrawn, the sample itself and the results derived exclusively from it should be destroyed'.

93 Fundamental Principles, supra note 6, Explanatory Notes, Principle 10, (With-drawal of Informed Consent).

94 Ethical Guidelines (as amended), supra note 8, Chapter 4 (Handling of Samples), 14 (Preservation and Disposal of Samples, etc.) (3).

95 Ethical Guidelines (as amended) supra note 8, Chapter 3 (Basic Attitude to Participants), 10 (Disclosure of Genetic Information) (10) (c).

96 UK Biobank Ethics and Governance Framework, Version 2.0 (July 2006), supra note 3, section I.B.6.

97 Icelandic Supreme Court, No.151/2003.

98 For a critique of this rationale, see Gertz, R. (2004) 'An Analysis of the Icelandic Supreme Court Judgement on the Health Sector Database Act', *SCRIPT-ed* 1(2): 241–258, arguing that the reasoning employed by the Supreme Court may lead to other complicated questions, such as whether an individual would also have legal standing to object to their living parent's data being included in the database.

99 Fundamental Principles, supra note 6, Principle 4 (Participants, and their Families and Blood Relatives).

100 Ethical Guidelines (as amended), supra note 8, Chapter 3, (Basic Attitude to Participants), 10 (Informed Consent) (9).

101 The representative can be selected from amongst the dead participant's spouse, adult offspring, parent, adult sibling or grandchild, relative living together, or a person considered to have been equivalently close to the dead participant.

102 Ethical Guidelines (as amended), supra note 8, Chapter 3, (Basic Attitude to Participants), 10 (Informed Consent) (10).

103 Ibid.

104 Stuart, supra note 19 at para. 1.6.

105 'Dispute over Data Privacy Halts Cancer Study' (2003) *Nature* 424: 359.

106 Fundamental Principles, supra note 6, Explanatory Notes, Principle 11 (The Protec-tion and Control of Genetic Information and Establishment of a Control System).

107 Ibid.

108 Ethical Guidelines (as amended), supra note 8, Chapter 2 (Responsibilities of Researchers), 5 (Basic Responsibilities of all Researchers, etc).
109 Fundamental Principles, supra note 6, Principle 12 (Leakage of personal information) 3.
110 Ethical Guidelines (as amended), supra note 8, Chapter 2 (Responsibilities of Researchers etc.), 7 (Responsibilities of the Chief Researcher) (7).
111 Stuart, supra note 19 at 7.10.
112 Ibid. Stuart states that all MHLW national research institutes are required to follow this procedure.
113 Masui and Takada, supra note 13 at 114.
114 Ibid.
115 Ethical Guidelines (as amended), supra note 8, Chapter 2 (Responsibilities of Researchers etc.), 7 (Responsibilities of the Chief Researcher) (7–8). Unanonymised samples and genetic information may also be provided to an 'outside institution' or transferred to subcontractors if the consent of the participant and the approval of the research institution and the Ethics Review Committee are obtained.
116 Fundamental Principles, supra note 6, Principle 15 (Disclosure of Information to Blood Relatives).
117 Ethical Guidelines (as amended), supra note 8, Chapter 3 (Basic Attitude to Participants), 11 (Disclosure of Genetic Information) (3-Bylaw 4).
118 Ibid.
119 Supra note 90.
120 Supra note 12.
121 Ibid.
122 The JMA's comments are indicative of the current distrust between physicians and clinical researchers in Japan. In his analysis of the JMA's motivations, Masui, ibid, has stated that he believes that the JMA is attempting to use this particular incident to exert political pressure on the relevant ministries and clinical researchers in order to 'win a greater say for physicians in planning and reviewing such projects'.
123 Ethical Guidelines (2001), supra note 7, Chapter 2 (Responsibilities of Researchers, etc.), 4 (Responsibilities of the Head of a Research Institution), (3- Bylaw).
124 Ethical Guidelines (as amended), supra note 8, Chapter 2 (Responsibilities of Researchers, etc.), 6 (Responsibilities of the Head of a Research Institution), (6); those entrusted with handling information relating to genetic research should be supervised appropriately by the head of the research institution, which may entail stipulating the procedures for protecting privacy within their employment contract.
125 Shirai, Y. (2003) 'The Status of Ethics Committees in Japan', *Eubios Journal of Asian and International Bioethics* 13: 130–134.
126 UK Biobank Ethics and Governance Framework: Summary of Comments on Version 1.0 (May 2004), supra note 63 at 4.
127 See Stuart, supra note 19 at 1.9.
128 Masui and Takada, supra note 13 at 117.

4 Trust and the creation of biobanks

Biobanking in Japan and the UK

Tohru Masui

Recent developments in biology and biomedical science have turned human beings into desirable research subjects. The revision to the Declaration of Helsinki in 2000 (WMA 2000) recognises the use of human material and data in medical research. Now, biomedical science has developed to the point that indirect medical research is partly replacing clinical research with living human subjects (HGC 2002; Masui 2002a, 2004a; NBAC 2001; NCBE 1995; Umeda *et al.* 1998). Moreover, economic interests in human health and disease augment this situation. Commercial enterprises are motivated to develop analytical agents, equipment, and information systems for indirect medical research. This situation is especially strengthened by the use of human genome information in biological studies of our species to stratify the human population into comparable and biologically standardised groups.

The indirect research methods using genome information, together with the use of biomarkers, medical records, and lifestyle data, are powerful strategies for understanding the human body and disease (HGC 2002; NBAC 2001; WMA 2000) and have increased the value of genetic material and data. For this reason, the establishment of biobanks (i.e. collections of genetic material and information) is a natural consequence of the developments in medical science.

However, many people are wary of the financial motivation of businesses involved in medical research, and this wariness could hamper participation in clinical studies. Since major operations such as large-scale biobanks require huge amounts of financial resources, and are under pressure to produce outcomes and return benefits, cooperation with commercial enterprises is necessary. Though commercial activity is regarded as the essential machinery of bridging a research outcome to a benefit for patients, commercial transfer of human information and material could cause intense scepticism among the public. The criticism and partial failure of the Icelandic Health Sector Database planned by deCODE Genetics, a commercial company, and the Icelandic state demonstrated difficulties in adjusting the interests and doubts among stakeholders (Annas 2000; Masui 2002b; Rose 2006).

People are more aware of their value to medical research, and they are increasingly conscious that their genetic material and data for medical research could be tightly bound to financial interests and commercial activities. Since genetic material contains individual genome information, which is both distinct and shared among human populations, and the stratification of human populations using genome information is critical for medical science, these two factors make genetic material sensitive.

Consequently, this chapter focuses on genomic research-oriented collections, such as biobanks, and some socio-political issues linked to such collections. I will also discuss the issues of benefit sharing, risk and benefit balancing, and trust and the motivation for donation. These issues will be examined in light of sustainable long-term policies by comparing biobanks in Japan and the UK. It will become clear that although Japan and the UK share problems, they solve them with different strategies, reflecting the diverging context of these countries.

The impact of developments in biology and biomedical science on medical research

The World Medical Association's (WMA) Declaration of Helsinki played an important role in the development of modern medical research. Human experiments during the Second World War made people aware of the need for appropriate guidelines on human experimentation. At the time, concerns focused on experimentation on the human body. The Nuremberg Code (1947) and the Declaration of Helsinki (1964) (WMA 2000) addressed these concerns. The Nuremberg code stands rigidly on the principle of consent; the Declaration tried to achieve practical adjustments. The Declaration follows the development of medical research, and the changes in the titles of the Declaration of Helsinki over the years clearly show that the targets of medical research have moved from individual patients to populations and from living human beings to genetic material and data (Table 4.1).

Table 4.1 Changes in the titles of the Declaration of Helsinki

1964: 'Clinical research on a human being'
  I.   Introduction
  II. Clinical research combined with professional care
  III. Non-therapeutic clinical research

1975: 'Biomedical research involving human subjects'
  Introduction
  I.   Basic principles
  II. Medical research with professional care (Clinical research)
  III. Non-therapeutic biomedical research involving human subjects
      (Non-clinical biomedical research)

2000: 'Medical research involving human subjects'
  A. Introduction
  B. Basic principles for all medical research
  C. Additional principles for medical research combined with medical care

In the 1964 version of the Declaration of Helsinki, the Declaration focused on traditional clinical research with physicians and individual patients. Here, scientific deduction was made on the basis of the accumulation of data from individual clinical cases. However, the 1975 version incorporated changes reflected in two instances of replacing vocabulary, i.e. from 'clinical' to 'biomedical' and from 'a human being' to 'human subjects'. These changes illustrate a development in biomedical science of an approach that emphasised the use of analytical technologies of the human body and diseases, that is, the biological view of the human body and diseases in medical research. For example, the biochemical analysis of blood with biomarkers represented an important advance in basic biochemistry and was soon widely used in a clinical setting. This advance pushed the development of modern medical practice based on a new scientific view of the human body and disease.

From the end of the 1980s to the 1990s, the Human Genome Project started slowly as an international collaboration. However, in 1998, the participation by Celera Genomics, a private company, threatened the status of the international collaboration. Around the same time, Iceland's parliament started to discuss legislation relating to the activities of deCODE Genetics, also a private company, in relation to the Icelandic biobank, and it passed the Icelandic Health Sector Database Act 1998. However, the planned project was unsuccessful, because the opt-out setting of participation and the right to withdraw the data had raised ethical and legal discussions. These incidents had a major impact on medical professionals and researchers and established a physical base or biological view of human beings.

The year 2000 Declaration of Helsinki, compared with the 1975 version ('Biomedical research involving human subjects') deleted 'Bio' from its title, changing it to 'Medical research involving human subjects'. The change illustrates that the majority of medical research is now based on a biological view of the human body and disease. In addition, the declaration defined *medical research* as involving 'identifiable human material and identifiable data'. The change in the definition meant that the WMA recognised the value of indirect methods of studying and measuring the human body and disease.

There might be also a political reason for the revision. In the 2000 term, Dr Eitaka Tsuboi, president of the Japanese Medical Association (JMA) at that time, was also the president of the WMA. A delegate from the JMA, who had joined the discussion, asserted during an interview that 'since so many non-medical researchers have come into the area of medical research using human material and data, the JMA and the WMA felt that it was our duty to protect the initiatives of medical doctors in medical research'.[1] As a matter of course, the medical doctor retains the initiative in clinical research involving human subjects. However, within the broad scope of 'indirect medical research methods', cooperation between medical doctors and other specialists has become a key issue.

Demand for biobanks

In the Declaration of Helsinki 2000 (WMA 2000), section four states that 'Medical progress is based on research, which ultimately must rest in part on experimentation involving human subjects'. Studies of the human body and disease are closely related to medical practice, and medical research is closely bound to the flow of money created by medical practice and care. The changes described in the previous section are related to the following two developments. On the one hand, the Human Genome Project paved the way for the construction of databases of the human body and disease, and made genome information the standard criterion for stratifying human populations. The availability of scientific criteria, i.e. genome information, transformed human beings into standardised and comparable subjects for medical research. On the other hand, studies of the human body and disease have become so closely intertwined with economic activities that the incentive for business to conduct medical research and participate in related activities has increased significantly. Both developments naturally require the collection of human material and information for indirect medical research in which non-medical research professionals from emerging research fields have their own identity.

This well-recognised environment for medical research means that human beings remain an attractive and appropriate subject for scientific studies. And biobank arrangements seem to function as bridging machinery used by medical doctors and non-medical researchers alike in biomedicine.

Biobank typologies

A biobank is a resource for indirect medical research on human health and disease. However, the term 'biobank' means something different depending on the context in which it is used. In this chapter, I prefer the superficial definition of biobank:

> A biobank is an organised collection of human material and information for medical study based on the optimal use of human genome information.

Biobanks can be private or public, commercial or non-profit based, large or small, etc. GeneWatchUK, a UK-based body monitoring research and policy issues of human genome research and genetically modified crops, defined biobanks as follows:

> Biobanks contain blood or cell samples from large numbers of people. Genetic information from each sample is linked to the individual's medical history and lifestyle data. There are already many small biobanks in the UK, and there are plans for a national biobank called UK Biobank. This would involve samples from 500,000 people and might be expanded later to include almost the whole UK population (GeneWatchUK 2006).

Any sample-based medical study can create biobanks. Within this broad category, however, we can classify biobanks by other criteria, i.e. target population, timeframe, and the health status of the population.

At the Organisation for Economic Co-operation and Development (OECD) meeting for 'Human Genetic Research Databases (HGRD): Issues of Privacy and Security' in Tokyo in February 2004, Dr Bartha Knoppers classified biobanks depending on the target population. There are basically four types: regional- or province-based biobanks, biobanks based on national populations, ethnic-based databases, and biobanks based on international/regional endeavours. Using a timeframe is another way of classifying biobanks. The biobank can be fixed at one point in time (cross-sectional), or it can include a follow-up research programme (longitudinal or cohort). Finally, biobanks can also focus on the health status of the target populations.

The OECD meeting used HGRD as an equivalent term for 'biobank'. The OECD report contains much information on the different types of biobanks in a global context (OECD 2006). In this chapter, however, I will discuss only the domestic issues of national biobanks.

Public and private biobanks

We can classify human resource banking systems into private and public ones. 'Public' here does not mean that the banking system is government-funded. Rather, 'public' refers to the function of biobanks in medical research. If a biobank distributes its collection to qualified researchers using an open and fair evaluation system, we call the biobank 'public'.

A 'private' biobanking system belongs to a specific organisation or person and supports specific researchers. Its biobanking policies govern the function of the organisation. The collection is not open to the public, and distribution of material and information is only for a collaborative or commercial purpose. Private biobanks can be supported by public funds. Alternatively, a publicly funded research project can create a private biobank, not open to qualified researchers.

During the OECD meeting in Tokyo, the following issues were discussed. When using genetic material and information, it is important for the biobank to be *public*, because (1) health data form part of a national heritage, (2) genetic data are essentially public and shared, (3) health data are recorded at public expense, and therefore, constitute a national resource, (4) it is essential that this resource remains open for public research, and (5) a public biobank avoids duplication of a collection. On the other hand, arguments for building a *private* biobank include: (1) health data are a commodity; (2) there is no inherent difference with traditional clinical trial data (data owned by the sponsor of clinical trials, data embodied in a drug master file for legislative purposes, and data exclusivity); (3) *de novo* collections require additional funds; and (4) the 'market place is always right' (OECD 2006). In addition it could be argued that a private biobank could serve the protection of sample donors well, as a private banking system could tightly control the use of human material and information. In practice, a public banking system would not be able to limit the use of data efficiently.

In the context of this discussion, even deCODE Genetics, which is a representative private biobank, was discussed in Iceland in a 'public' light:

> Due to the nature of the data and their origin, they cannot be subject to ownership in the usual sense, neither by institutions, companies nor individuals. It is, however, both fair and a duty to utilise the data in the interests of the health sciences and public health. This can best be done by the government authorising the creation and operation of a single centralised database, which has the following benefits: (1) acquisition of new knowledge of health or disease, (2) improved quality and economy in the health system, (3) development of high-technology industry and employment in Iceland, and (4) potential for attracting business to Iceland (OECD 2006).

Therefore, deCODE Genetics works under the guidance of the government and represents national benefits even though it initiated the commercial operation (Masui 2002b; Rose 2006). In this sense, the deCODE case is a quasi-public activity. This discussion clearly demonstrates the public nature of human genome material and data.

In any case, a biobank should maintain a public policy over data and sample collection. Another reason for public concern over biobanks lies in the way they are used and whom they benefit. It is important for the users of a biobank to maintain their public status and act accordingly. This leads us to the issue of the need for benefit sharing, though this discussion, as I will explain below, has great difficulty in resolving the problems of maintaining anonymity and the incompatibility of the benefactor and the beneficiary. To complicate the discussion on the nature of 'public' even further, it has to be pointed out that the concept has different implications depending on the context. This point will become clear when comparing the Japanese and British biobanks in the latter part of this chapter.

There is another reason for not regarding biobanking as a private commercial activity. Research resource banking systems require a large amount of 'dead stock'. Dead stock refers to potentially useful stock of data and/or materials that are not necessarily in constant use. This is important, for if a biobank collects only currently useful 'valuable' stock, the biobank would be out of date as research trends shift. Collections should not be too narrow and must be prepared for possible changes in the future. In this respect, the biobank does not fit the requirements of a private commercial activity, because a commercial activity seeks efficiency and does not tolerate 'dead stock'. This discussion raises issues of funding for biobank systems associated with their role in society.

Japanese genome research

In 2000, the Japanese government started the Millennium Project, a five-year target for science and technology in seven commercially promising areas (Government Office 2006). The project involved research on the human genome and required the creation of guidelines.

Coinciding with the planning of the Millennium Project in 1999, three cases of genome analysis were reported without the specific informed consent of the donors (METI 2006). And because genetic/genome information in Japan is regarded as the ultimate form of private information (Nukaga and Tsutani 2006), the mass media were eager to report misconduct in genome research. The three cases helped the government and academia to establish a consensus for the need of official guidelines on genomic research. The first ethical guidelines on genome analysis were issued on 28 April 2000 (MHLW 2006). The Ministry of Health, Labour and Welfare (MHLW) published 'the Millennium Guidelines'(*Idenshi Kaiseki Kenkyu ni fuzuisuru Rinrimondai tou ni taiou surutame no Rinrishishin*) for the regulation of millennium genome projects.

Independent of the guidelines, the Bioethics Committee, the Council for Science and Technology (BC, CST) (*Kagaku Gijutsu Kaigi, Seimei Rinri Iinkai*) issued the 'Fundamental Principles of Research on the Human Genome' (*Hito Genomu Kenkyu nikansuru Kihongensoku*) (CST 2000), on 14 June 2000. 'The Millennium Guidelines' were the standard operational protocol for researchers. The principles were influenced by 'the Universal Declaration on the Human Genome and Human Rights' of UNESCO 1997 (UNESCO 1997). These principles focused on a conceptual framework for genome research and were not intended as practical guidelines for researchers.

Other ministries have been funding human genome research projects, so the government decided to establish the guidelines in collaboration with three ministries: the MHLW, the Ministry of Education, Culture, Sports and Technology (MEXT), and the Ministry of Economy, Trade and Industry (METI). 'The Ethical Guidelines for Analytical Research on Human Genome/ Genes' (*Hito Genomu/Idenshi Kaiseki Kenkyu ni kansuru Rinrishishin*) were issued on 29 March 2001 (MHLW, MEXT, METI 2001). The new guidelines replaced the old guidelines from the MHLW, merging the trends from the old guidelines, but covering a wider range and in some areas loosening the regulatory framework (issues of anonymity, surrogate decision-making process, and existing holdings).

From 2003, the Data Protection Bill (*Kojin Jyouhou Hogo Houan*) was discussed intensively, and the public became interested in the discussions regarding data protection. However, the use of genome information in research was not closely examined, which may have occurred because academic research was exempt from the Act (Article 50) and academic areas already had the guidelines. Of course, modifications to the guidelines were necessary according to the regulatory framework of the 2003 Act (*Kojin* 2003).

According to the Act, the MHLW drafted the 'Guidelines for Appropriate

Handling of Personal Data in Medical and Care Services' (*Iryou/Kaigo kankei Jigousha ni okeru Kojinjouhou no Tekisetsu na Toriatsukai no tameni*), issued on 24 December 2004 (MHLW 2004). The guidelines established due process for data processing in medical practice and care. However, the guidelines provided very little information about how to transfer medical information from the medical field to research.

Since the Data Protection Act (*Kojin* 2003) became effective on 1 April 2005, the guidelines for medical research had to be adjusted to comply with the Act. The revision process was completed in December 2004, and became effective in April 2005 (MHLW, MEXT, METI 2004). The revisions focused on the area of data protection and the principles of safeguarding data, i.e. systematic, personnel, physical, and technical safeguards, and other adjustments according to OECD data protection principles (OECD 1980) that are incorporated into the Act (*Kojin* 2003).

The government also tightened the regulatory consistency between the explanations of the research purposes when obtaining informed consent from sample donors and the actual use of materials and information in medical research. As the new guidelines require the assurance of explicit consent, the range of research applications after informed consent is limited. These requirements in general are not suitable for biobank activities, because biobanks, in principle, serve future, still unknown research purposes.

Genomic research has generated several biobanks in Japan. I will describe two of them: the Pharma SNP Consortium,[2] and Biobank Japan.[3]

The Pharma Single Nucleotide Polymorphism Project

The Pharma SNP Consortium (PSC) was established in September 2000, and started its collection in February 2001, just before the three ministries jointly issued the guidelines for analytical research on human genome/genes (MHLW, MEXT, METI 2001). The Consortium gathered forty-three pharmaceutical companies, and collected samples from about 1200 'common' Japanese people. It collected the blood of donors, blood test results and data from questionnaires. The samples were completely anonymous and processed as unlinked data sets of DNA, blood, and sample information. The project was conducted at the Tokyo Women's Medical University, under the leadership of Professor Naoyuki Kamatani, and Professor Yusuke Nakamura's group of researchers at Riken[4] analysed the single nucleotide polymorphisms (SNPs). The project analyzed the frequency of SNPs of about 170 genes of drug-metabolising enzymes in a population of ordinary Japanese. The results are expected to serve as the normal control for pharmaceutical research projects.

The samples were donated with informed consent, but with two conditions: the permission for use by the PSC and the guaranteed donation of the samples to a public bank after the completion of PSC's project. The people who agreed to these two conditions participated as volunteers. The PSC processed the donated blood not only to purify the DNA for its own project, but also to make immortalised cell lines by infecting the B-lymphocyte cell population of blood with the Epstein-Barr (EB) virus. Artificial infection of human B-lymphocyte with the EB virus causes an extended culture period or immortalisation without much alteration of individual genome integrity. Therefore, the EB virus transforms a limited resource from a donor to an almost unlimited resource of human DNA. After the completion of the project, the cell lines were donated to the Human Science Research Resource Bank,[5] and to the Japanese Collection of Research Biologicals, JCRB Cellbank.[6] These banks collaborate under the guidance of the MHLW.

Although this research plan received partial funding from the MHLW, it was initiated and primarily funded by consortium members of industry and was not part of the Millennium Project. In 2000, there were no specific guidelines applicable to the PSC genome project. Therefore, following publication of the Millennium Project Guidelines, the Consortium published its own guidelines.

This PSC genome project constitutes the first major collection of genome material for research by industry in Japan. As the collection of genome material and data was still a publicly sensitive matter, PSC maintained a cautious attitude during the whole period. The data was released in February 2004, and just after two months the data had been made accessible to the consortium members. As in other countries, the industrial initiative requires stricter standards than those of academia because journalists and the public watch their activities much more carefully than those of academics. The public policy was applied not only to the operation, but also to the collection of data, and the Consortium opens all its activities on its website. There may be other reasons for establishing public policy for PSC activities. It might be difficult to establish a consistent plan for the distribution of benefits among the members of the Consortium without it. Therefore, if a policy was devised for allocating certain periods of exclusive use of the data by members of the Consortium, the members would have to accept the policy of making the acquired data public. The other issue might be the continuation of funding. The generation of research data does not require much effort and funding after collection. On the other hand, the collected and immortalised materials demand care and back-up funding for long-term maintenance. The best use of publicly collected material might be considered as an ethical use of donated materials under the appropriate informed consent. This situation may have pushed the Consortium to establish its own public policy: all the established cell lines were donated to public cell banks. In this way, the collection became a useful tool in the public domain.

At the time of the donation of materials to the cell banks, the accompanying sample data from the cell lines remained at Tokyo Women's Medical University. In this way, the data and materials were managed independently, maintaining a high level of security. Professor Naoyuki Kamatani at the Women's Medical University was responsible for providing the available data upon request. The Consortium only lasted three years, from September 2000 to May 2003, but the collection has become a valuable public resource in Japan.

The data generated from the project is also accessible to the public. Those who wish to access PSC materials are asked to obtain authorisation for their project plans from their research ethics committee. This process is enforced because public policy stipulates the ethical justification for the use of donated materials and information. More importantly, donors were informed at the time of giving consent that a research ethics committee would authorise the future use of their materials from the public cell banks.

In the beginning, public biobanks discussed the international distribution of PSC materials. However, since many Japanese pharmaceutical companies are international, public biobanks decided to distribute the materials and information internationally as well as in Japan. In accordance with public policy lines, the PSC completely publicises its activities, protocols, proceedings of research ethics committees, research results, and materials and information. PSC, however, is an exceptional case in Japanese research, especially the 'public status' of the bank and its political transparency.

Biobank Japan

In the official Japanese papers, the concept of the 'biobank plan' appeared in a report from the Subcommittee on Science Project Evaluation and Life Science, Genome Research Working Party (*Genomu Kenkyu Ryouiki Shou-Iinkai*) on 20 March 2002.[7] The report included recommendations for a policy on genomic research in Japan. The report defined the 'biobank' as follows:

> a government supported collection of, preferably, immortalised and propagable EB virus transformed B-cell lines derived from patients and their family members, patients that have experienced adverse drug reactions, and volunteers. The collection should be voluntarily initiated and organized by researchers, medical doctors, and the pharmaceutical industry. The samples should be maintained and managed by the biobank and provided to qualified researchers upon request [translated by the author].

This description of the biobank reflects the practice of PSC described above. At this stage, the report recommended the collection of samples from 360,000 people, including 20,000 'normal' volunteers.

After the publication of the report, the Evaluation Subcommittee for Research Planning (*Kenkyu Keikaku Hyouka Bunkakai*), a subordinate of the Science, Technology and Academia Committee of the MEXT, issued the 'Promotion Strategy of Research and Development of Life Science' (MEXT 2002). In the report, the subcommittee recommended 'establishing the infrastructure for the management of resources and technology to realise personalised medicine (tailor-made medicine)' [translated by author]. On 24 July 2002, the Cabinet Committee of Science and Technology (*Sougou Kagaku Gijutsu Kaigi*) issued the 'Estimation of Science and Technology related Budget in 2003' (*Heisei 15 nendo no kagakugijutsu yosan no gaisanyoukyu ni mukete*). The suggested budget was settled for the next fiscal year under the project title, 'The Realization of Tailor-made Medicine on the Genetic Characteristics of Individuals' (*Kojin no identokusei ni oujita teirameido iryou no jitsugen*), earmarking an estimated 40 billion yen (US$330 million) over five years (in a recent presentation, US$180million, October 2006). This plan became known as the Biobank Japan Project.

Biobank Japan aims to collect data from 300,000 patients (or cases) with over forty major diseases, in cooperation with over forty cooperating hospitals (Ohnishi and Nakamura 2005). This is primarily a five-year research project for collecting samples and information, with the supplementary function as a biobank. The study design includes the preparation of controls from the samples of patients with other diseases. The project defines the biobank as:

> A collection of DNA and serum of patients with certain diseases and patients with effectual and adverse drug reactions. Its infrastructure facilitates the analysis of SNPs and the creation of order-made ('made to order'; 'personalised') medicine [translated by the author].

The Biobank Japan Project started in the fiscal year of 2003 with the project title 'Realization of Order-made Medicine (*Ooda-ameido Iryou Jitsugen ka Purojekuto*)'.

Although the term 'order-made' is not English, its meaning is significant to the project. Professor Nakamura, its leader, favours the use of the term 'order-made', though the documents from the MEXT stick to the 'tailor-made' medicine. Professor Nakamura expressed his view that the concept of 'tailor-made' smacks of a class society, and said that genome science should provide benefits to the general public, not to a privileged elite. Therefore, if the research outcome only benefits the wealthy or is distributed along class lines, it would not be consistent with what Biobank Japan intends to offer the public. The spirit in which Biobank Japan was established, then, would have to be in agreement with the purpose of the public medical insurance system in Japan, which to a great extent has accomplished equality and solidarity with the financially underprivileged.

Biobank Japan is a large, case-control study and does not intend to perform follow-up studies. The project compares the genome–environment interactions of patients with a certain disease with those of carriers of another disease. For example, patients with diabetes might be studied in comparison with those with cancer. In addition, the secondary use of the collected material and information is part of the project plan. Biobank Japan announced that it provides collected material to qualified researchers, who have authorisation from the ethics committees of their institutes and meet the required qualifications of the biobank. However, Biobank Japan provides the DNA samples, but not the data, to the researcher. In the original plan, the research project which had received DNA material would provide the analytical results of the DNA to the biobank, and the data analysis centre of the biobank would analyse the DNA data with other medical and lifestyle data and return the final results. As I will describe below, the governance of material and information by Biobank Japan is very different from that of UK Biobank, which plans to provide data, but not distribute DNA material.

The British plan for genome research[8]

The UK House of Commons' Science and Technology Committee issued the key report 'Human Genetics: Science and Its Consequences' in 1995. This report provided the starting point for research policies on genome science and its application in the UK. Although, the committee criticised UK Biobank in its report 'The Work of the Medical Research Council [MRC]' in 2003 (House of Commons 2003), the 1995 report paved the way for the establishment of UK Biobank.

In 1998, two important events occurred in the field of genomic research. One was the participation of Celera Genomics in the human genome project, and the other was that the Icelandic parliament passed the Icelandic Health Sector Database Act. Although the Icelandic Health Sector Database was not successful for various reasons as I mentioned earlier, the Icelandic controversy stimulated discussion on the value of human genetic material and data.

In the planning process of UK Biobank, the MRC at the same time discussed a policy for the governance of materials and produced the interim guidance document 'Human Tissue and Biological Samples for Use in Research – Operational and Ethical Guidelines' 2001 (MRC 2001). An important point of these guidelines was to discuss the public status of human tissue collections and the governance of the funding body handling the collection.

UK Biobank

The author has been following the developments of UK Biobank from its inception (Masui 2002b, 2003, 2004b; Masui and Takada 2005). It is clear that UK Biobank cannot be regarded as a research project *per se*. Its aim is to collect biomaterials (blood and urine) and information (medical information, lifestyle questionnaires, and measurements) from 500,000 UK citizens, ranging in age from 40 to 69 years old, and to conduct a follow-up in 20–30 years.[9] Its aim is:

> UK Biobank project will enable scientists to gain a unique insight into the genetic and environmental causal factors associated with a wide range of debilitating diseases, providing vital information needed to work on future preventative and curative measures (Wellcome Trust 2006a).

The UK Biobank project is expected to provide research resources to qualified researchers. It is conceivable that the core concept of the UK Biobank project is that of a social experiment set up to obtain the best possible outcome from human genome research in the context of the medical health sector in the UK. For this reason, UK Biobank serves as a custodian of the resources of UK citizens.

The UK started discussions on a population biomedical collection in late 1998. In May 1999, the MRC and Wellcome Trust (WT) held a workshop entitled 'UK Population Biomedical Collections', at which it reviewed major genome cohort projects funded by them. These funding bodies jointly announced that they would fund the preparatory steps of the UK Population Biomedical Collection in June 1999. In October and November of the same year, the WT held two consecutive meetings, the 'Workshop on Pharmacogenetics' and the 'Workshop on Human Biomedical Collections'. Moreover, in November 1999, the MRC issued the interim guidance entitled 'Human Tissue and Biological Samples for Use in Research'. These activities followed and strengthened the joint announcement in June. In the ethical guidelines, the MRC proposed a material governance policy and discussed public collection and custodianship because the fundamental question of a large biobank concerns its impact on not only individuals, but also society at large. For this reason, the MRC and the WT have maintained a policy of engagement in public debate on the public nature of the collection. These discussions focus in particular on the collection of UK Biobank and the National DNA banking network (Rawle 2003).

During the preparatory stage of UK Biobank, the MRC and the WT surveyed public perspectives on genome research and biobanks. They attached great value to obtaining public trust to facilitate progress in the biobank (PSP 2002). The acceptance rate for enrolment was estimated at around a quarter or less. It was therefore important for the success of UK Biobank that people in the locations for recruitment and in that particular age range were favourably aware of the aims and the process of the project. This meant that UK Biobank attached value to its public status and was motivated to survey the public perception of UK Biobank (PSP 2002). At the same time, they studied the perceptions of professionals in UK Biobank (Hapgood *et al.* 2006). Even though there was continuing criticism of the UK Biobank project (Giles 2006), the funding bodies and the biobank have come to feel confident and comfortable with its status.

In the spring of 2003, the funding bodies, consisting of the MRC, the WT, and the Department of Health, appointed a CEO and started practical moves for the project. UK Biobank was registered as a limited company and as a charity. In 2005, a new CEO was appointed. In my interview with Professor Rory Collins, the CEO of UK Biobank (February 2006), Professor Collins said that the project was planning to recruit people independently from general practitioners, and that UK Biobank had applied to the Patient Information Advisory Group (PIAG),[10] and obtained authorisation from the Minister of Health to survey potential participants without their consent. The terms of the PIAG are as follows:

Section 60 of the Health and Social Care Act 2001 provides the power to ensure that identifiable patient information needed to support essential NHS activity can be used without the consent of patients. The power can only be used to support medical purposes that are in the interests of patients or the wider public, where consent is not a practicable alternative and where anonymous information will not suffice.[10]

In short, although UK Biobank itself does not have any specific legal powers, its activities are framed and supported by the Data Protection Act of 1998 and the Human Tissue Act of 2004.

Differences between Japanese and UK biobanks

As mentioned above, definitions of biobanks differ in a variety of respects. Although the biobanks in Japan and the UK are planned to be public, their different conceptualisations of public, risks, benefit, and trust affect their 'public' status. The scope of the differences reveals important points of consideration in the concept of collecting human material and information.

Preparing Biobank Japan and UK Biobank

The implementation of the plans for Biobank Japan started in the fiscal year of 2003. Until then, the Millennium genome studies (which had commenced in 2001) had experienced two years of practice, during which its guidelines had already been revised once in 2001 (MLHW, MEXT, METI 2001). The planning committee for the project believed that Japan was prepared for Biobank Japan, and that there would be no great need for further preparation (Ohnishi and Nakamura 2005). In fact, the project was believed to be a large and expensive version of the Millennium Project. The Japanese government did not regard this project as something new, so it had no notion of having to deal with the particular challenges of a social experiment.

The PSC project had been organised as an industrial initiative and had collected about 1200 blood samples from 'ordinary' Japanese people. It had immortalised the blood samples and donated them to the public banks. As described earlier, the success of the PSC project may have had a strong impact on Biobank Japan, because the practice of PSC had influenced the definition of 'biobank' used in the March 2002 report 'Genome Research Working Party' (*Genomu Kenkyu Ryouiki Shou-Iinkai*), chaired by Dr Yoshiyuki Sakaki. Biobank Japan established the Ethical, Legal, and Social Issues (ELSI) Working Group in 2003. And in 2004, the MEXT, the funder of the bank, established an ELSI Committee independently from the project.[11] Its task was to monitor and to advise the biobank project.

For UK Biobank, the draft of the Ethics and Governance Framework was open to public comments from September 2003, and public comments were summarised and published in May 2004. Then, the Ethics and Governance

Committee (EGC) first met in November 2004.[12] They discussed the issues of supervision and ethics and governance of UK Biobank. Preparatory activities for UK Biobank differed substantially from those of Biobank Japan, especially regarding the relationship between biobank and society. In 1995, the Science and Technology Committee of the House of Commons reported its views in 'Human Genetics: Science and its Consequences'. And in 1999, the Human Genetics Commission (HGC) was established to create long-term science and ethical policies on human genomic research.[13] Until the establishment of the Committee, there had been no initiative to create long-term policy recommendations. Since creating and implementing such a policy for genome research would require the cooperation of multiple ministerial departments, a central committee was regarded as indispensable.

In addition to this top-down infrastructure, the Department of Health and the Department of Trade and Industry decided to jointly fund the Genetics Knowledge Parks (GKP) for five years, from 2002 until 2007.[14] This created a bottom-up movement in the field. In 2004, the Human Tissue Act 2004 was passed,[15] and in 2005 the Human Tissue Authority was established to address the issues of human tissue,[16] the essential resource of genome information. This would provide the legal framework for the use of human tissue in genome research as a whole.

These elements of the policy of the House of Commons and the HGC, the engagement of GKP and the Ethics and Governance Committee (EGC, see below) and the legal framework, together with UK Biobank, give an edge to the UK in the area of human genome research. Moreover, these movements indicate that the UK government has seriously considered indirect human experimentation. This may be due partly to the UK tradition of epidemiology, which involves collecting medical data from UK citizens. The use of genome information is regarded as central to its future success. The UK policy, then, was a determination to build a robust trust based upon the dialogue between stakeholders. These activities were regarded as necessary for a mature society to consider and accept the possible outcomes of genome research.

Differences in the origin of researchers in UK and Japan

The academic background of researchers in biobank projects is an important factor in their development. In Japan, genetic research of single-gene disorders started in the 1980s. These studies were successful, and their outcomes created excitement among citizens and funding bodies. For this reason, genetic research based on DNA science has acquired a good reputation and much funding. The conclusion of the Human Genome Project changed the interests in genetics research from single-gene disorders to genome research of multifactorial diseases. In Japan, genetic researchers of DNA science who had been working on single-gene disorders were the first to move into the area of biobanks.

In comparison, in the UK, it was epidemiologists, by incorporating genome information into their epidemiological analytical processes, who have led UK Biobank and existing genome epidemiology projects. Genome information would be a powerful tool in stratifying human populations, but the contribution of genetic factors is much smaller in common multifactorial diseases. Careful handling and examination of other human environmental information is essential to the study of multifactorial disorders of the human body. Epidemiology, which traditionally pursued the association of diseases with various biological and environmental factors, also provides methods for the study of multifactorial diseases.

The study of single-gene disorders has an exceptional position in the study of human diseases, focusing on the causative relationship between genetic information and disease. However, epidemiologists have struggled to obtain comparable data from multifactorial phenomena. In multifactorial research, a genomic factor does not necessarily constitute a major factor in causing a common disease. The study of the interaction between the genome and the environment requires skill in the collection of non-genomic information. In this sense, epidemiologists have an advantage in their academic training of collecting information on the human body in a comparative context.

Obtaining disease data is not without its complications, because individual doctors have different styles of diagnosis, and to study human disease requires standardisation of the description of each disease. Researchers then require standardised data on patients. Thus, data from blood tests seem suitable and useful, but they show only a one-time section of a patient and have limited value.

Therefore, obtaining human data may benefit from the experience of epidemiologists. Due to the difference in academic origins of biobanking in Japan and in the UK, then, the epistemological backgrounds of the biobanks in Japan and the UK are quite different.

Regulation of sampling

In Japan, clinical technicians draw most of the blood samples. However, a clinical technician can only draw blood up to 20 ml under a MHLW Notice.[17] PSC collected 30 ml (10 ml for biochemical analysis, 10 ml for DNA isolation, 10 ml for immortalisation of B-cell fraction), and Biobank Japan collects 20 ml (6 ml for biomedical analysis, 7 ml for DNA and plasma isolation, and 7 ml for serum isolation). The 20 ml limitation of the MHLW's Notice may seem to set a limit for sample collection.

There is another limiting factor in Japan. In ethical discussions, informed consent is of primary value in medical research planning. In Japan, the philosophy of informed consent strictly regulates the amount that can be sampled and the use to which it may be put. Consequently, medical researchers are very careful not to take 'unnecessary' or 'excess' blood when sampling. This trend may also limit the sample size.

Ten millilitres of blood yields about 200 µg of DNA. Thus, in Japan, the immortalisation of samples is favoured because the immortalised cell line produces unlimited amounts of DNA samples from each individual. So the biobank report included an immortalisation plan.[18] However, the immortalisation process for blood is expensive and causes an increase in the chance of cross-contamination of samples. Therefore, the immortalisation of samples requires great care and a large budget.

As for sampling regulation in the UK, UK Biobank issued 'Sample handling and the storage, subgroup protocol and recommendations, version 1.0' on 7 July 2004 (UK Biobank 2004). It stated that the Biobank collects 40 ml blood (30 ml for DNA and plasma, 10 ml for serum). According to this sampling schedule, UK Biobank takes more than four times the samples for DNA analysis compared to Biobank Japan. Moreover, UK Biobank seems to be planning to save a limited amount of samples by conducting genome analysis within the biobank. In this way, they can recover leftover samples from the genome analysis. However, Biobank Japan has decided to provide a 5 µg aliquot of DNA and a 0.5 ml aliquot of serum from individual samples to a qualified researcher.[19] This illustrates a different strategy in governing samples.

On the distribution policy of data samples, UK Biobank plans to provide sample information to researchers, but Biobank Japan does not give the sample information to the researcher. In summary, the biobanks of Japan and the UK have greatly contrasting policies for governing genetic material and information.

Enrolment focus – differences in follow-up and risk management[20]

The UK Biobank project is a cohort study for up to thirty years, and enrolment will focus on 'ordinary' UK citizens. The bank follows the health status of participants during this period. The considerable follow-up process requires referencing and collecting samples and data from individuals. This requires long-term collaboration between researchers and participants. Although, in the original plan, the biobank itself does not perform the research, custodianship requires a well-prepared strategy and structure for supporting a dialogue among stakeholders in cooperation with the EGC.

In contrast, Biobank Japan aims to collect samples and information from patients, and perform research of their own, and does not include a follow-up programme. Patients are generally more amenable to research enrolment than volunteers. In the stage of research planning, one disease is compared to another disease as the control group. The collection schedule for Biobank Japan is of lower risk than that of UK Biobank because it does not contain long-term collaboration with research subjects and society.

The various scandals of the 1990s in the UK may have stimulated careful risk management of the biobank planning process. In particular, the public sector had lost credibility after scandals concerning BSE, GMO, and human tissue retention without appropriate consent. Programme officers from UK Biobank and the funding bodies were very careful about publicity, until the funding bodies officially announced the establishment of the UK Biobank project in April 2001. Though the project requires enthusiastic participants, expectations that are too high may mislead people and result in the collapse of the project.

The idea was that 'too high of an expectation kills all'. Until they were ready to respond to the public or opponents, they wanted to keep UK Biobank low profile. Another aspect to the UK risk management strategy concerns the nature of the benefit of UK Biobank. The programme officers stated that the benefit of UK Biobank lies in the increase in knowledge. This basic statement is essential because it shows UK Biobank's intention to support competitive research projects without making any promises in principle. However, this is a difficult task.

If researchers only raise 'appropriate' expectations, they may not obtain sufficient support from the non-expert population, and funding bodies might lose interest in their projects. Therefore, researchers tend to exaggerate the promises of their research programme. Programme officers of UK Biobank, during my interviews with them, often complained about researchers who are promising too much. It is conceivable that the researchers felt that they should obtain the support of citizens and motivate funding bodies in the biobank project. This might be the cause for the behaviour of researchers during the incubation period of a research programme.

While transferring the knowledge on hypothetical and uncertain scientific research to non-experts, the possible failure of experimentation should be made explicit and discussed rationally and calmly. However, research results from studies in scientific communication revealed that the transfer of the hypothetical nature of science knowledge to non-scientists and even to scientists is not unproblematic (Shamos 1995).

Biobank Japan uses a very different risk management strategy. Biobank Japan operates in the name of the 'Realisation of "order-made" medicine', and it emphasises the practical benefits from the beginning. In Japan, among scientists there are many complaints about the poor acceptance of clinical research. There are various reasons for this, though the phenomenon is regarded as a reflection of a poor understanding of human experimentation. Next, it reflects the Japanese misunderstanding of science, i.e. science and technology are seen as the tamed servants for the well-being of humans. Emphasising the practical benefits of a research project is important for obtaining funds and the support of citizens, but this strategy increases the risks of a misunderstanding of the nature of science. In order to avoid the risks, the projects must guarantee substantial outcomes. It is said that this sequence of providing prospects beneficial to society is important in promoting research and it was praised in the third 'Science and Technology Basic Plan (*Kagaku Gijutsu Kihon Keikaku*)' of the Council for Science and Technology Policy.[21] However, this positive-reward cycle, i.e. positive outcome cultivates support of science, cannot absorb the uncertainty and hypothetical nature of scientific research, and might increase the risks associated with medical research and create a poor understanding of science.

Balancing risk and benefit

In medical practice, a patient and a doctor focus on the risks and benefits of a treatment to the patient. In the clinical setting, then, the subject is singular, and the focus is on the improvement of the condition of the patient, i.e. the well-being of the patient is the primary concern. By contrast, in the context of biomedical research involving human subjects, medical doctors and researchers cannot focus only on the benefits to a particular participant. The situation is described in the Declaration of Helsinki section 7, 'In current medical practice and in medical research, most prophylactic, diagnostic and therapeutic procedures involve risks and burdens'. Of course, the participant's welfare is the primary concern in research planning. Therefore, the Declaration states that: 'In medical research on human subjects, considerations related to the well-being of the human subject should take precedence over the interests of science and society (section 5)'. However, medical research is also concerned with the due course of scientific human experimentation. Since medical research needs to obtain scientific and comparable results, the experiments require randomised trials and placebos. The World Medical Association has spent much time discussing the 'placebo-controlled trial' and incorporated notes of paragraph 29 and 30 in the Declaration. Certainly, following due procedure places 'risks and burdens' on the participants.

Medical research of this kind therefore basically aims to bridge the participants' current risks and the burdens and benefits of the next generation. This idea is a central point of departure in medical experimentation. Thus, if we take into account the differences in medical research on living human beings and indirect medical research using genetic material and data, the latter could substantially reduce the risks and burdens of a participant in medical research experimentation. It is imperative that indirect medical research supported by biobanks is done as efficiently as possible so as to reduce the risks and burdens on human beings. In this sense, biobanks could have indispensable value to medical development and to the welfare of research participants and should be promoted.

Though benefits to the public are also an important issue for biobanks, the UK and Japan take the same stance of not paying the donors for participation. Since medical research produces benefits that are not meant for the participants themselves, but for the next generation, the donors expect that their tissues, cells, and information will contribute to the health of future generations. The policy requires the biobank to have public status and responsibility.

If we think of benefit sharing as a form of reciprocity, the ethical burden lies with the researcher's use of human material and data. As medical research may result in commercial benefit, academic honours, and public reputation, researchers should be aware of the public status of research resources. What it means in reality varies in each case. However, researchers must pay considerable attention to donors and society, as their cooperation enables them to study human beings and to perform indirect human experimentation.

No direct personal benefit to participants

Both the biobanks in Japan and the UK are careful to deny the personal gain, not only monetary but medical, of individual participants. The public has difficulties in accepting this.

In Japan, the idea of order-made or personalised medicine has been very popular, since it was thought that genome research would lead to a miracle cure for the diseases of individual patients. This idea seems to derive from the images of the workings of antibiotics against infectious agents. Or sometimes participants misunderstood how the genome research could benefit their own health. However, it takes a long time for research outcomes to produce a practical cure.

The idea was put forward that 'personalised medicine' resembles traditional medical practice focusing on individual patients; diagnostic and therapeutic information created by genome research of multifactorial diseases would not work differently from traditional medical knowledge in the essential sense. It was stressed that research results of population studies could be adapted to an individual patient at the clinical phase. In these aspects, medical research and treatment are different.

These ideas should be familiar to the general public. However, ordinary citizens do not see them in a favourable light, since direct and substantial benefit is easy to understand. Therefore, research domains should explain the consequences of medical research to make the participants and public understand them.

It is interesting that UK Biobank suffered a similar problem concerning the benefit to participants. In the public comments on the Ethics and Governance Framework 2003 (UK Biobank 2003), there were a few queries on what could be regarded as a substantial benefit to participants. As a policy, UK Biobank clearly denied the benefit to participants. The Ethics and Governance Committee, at the first meeting (November 2004), took the initiative in denying any substantial benefit or data return to participants and reconfirmed the voluntary status of the participants, so as not to raise unrealistic expectations (EGC 2006).

Although the biobank seems to be a natural consequence of post-sequence biomedical science in the genome era, we are not sufficiently prepared yet to accept this type of 'indirectly invasive' activity. Only if we build on the basis of trust can the idea of a biobank survive the many problems to come. Since the idea of research is not a promise in the future, in some cases 'unconditional' support should be necessary.

What is the role of trust in the participation in medical research?

In his memorial essay on human experimentation, the philosopher Hans Jonas claimed that in the course of medical research we should not depend on 'trust' because trust is not based on an equal relationship and on independence (Jonas

1974). Therefore, the trust relationship in principle cannot call for voluntary donations.

His claim is fundamentally important and correct, but I believe that in medical research it would not be practical and in some sense wrong. There are never entirely equal relationships between medical doctors or researchers and patients or participants in medical research. Medical research now involves non-medical researchers and often involves commercial interests. In practice, the researchers propose a research project to patients or healthy volunteers and try to obtain motivated participation. Of course, the motivation of the researcher not only focuses on the public good, but also on their personal and research interests. In the dialogue with research participants, the research domain encourages them with openness and a sense of responsibility to understand the intended results of the medical research project, which are intrinsically uncertain and can only be revealed in the (sometimes faraway) future. Without a responsible attitude, the participants cannot be encouraged to support medical research in the long run. If a project cheats its participants, medical research generally loses credibility and cannot build a trustworthy partnership between researchers and participants. Though the described process seems essential, a project needs to collect a certain number of participants within a certain period and cannot often tolerate the burden. The situation is painful, therefore, and the professionals in medical research should develop norms on medical research.

In this sense, the informed consent process of participants is important to researchers. This process gives researchers a chance to reflect on what they are doing with and to the participants. The process is indispensable in developing a norm for medical researchers. We should try and seek to establish professionalism in medical research. Professionalism is traditionally established in medical, legal, and religious areas. I believe that in the era of human experimentation we need to consider the issue seriously. In doing so, the words of Richard Shryock might be helpful to understand and make up our minds to take the nature of medical science of human diseases and its difference from 'science' *per se* (Shryock 1974).

> Physicians were the only scientists who, because they were also practitioners of a vital art, were constantly being pushed to hasty and careless conclusions. Other research men, uncertain in the face of new problems, could suspend judgment and proceed with due caution. Practitioners confronted with dying patients did not dare to wait; they must act quickly and, if necessary, 'take chance.' Even during hours stolen for research, they were still under pressure to get practical results *as soon as possible*.

Shryock originally wrote this about 70 years ago, and the situation with current medical research does not seem to have changed much. We have to keep this in mind and maintain the dialogue between the various stakeholders and among medical researchers and doctors. In this way, we can judge ourselves better in the coming era of human experimentation.

Reflections

Though the biobank projects discussed in this chapter may have differences, a minimum requirement for all biobanks should be the acquisition and maintenance of public trust. How can medical research, including genome research, build or acquire trust? Among policy-makers and related people, it is generally believed now that public engagement is essential. However, it is far from clear if public engagement is truly succeeding in engaging the public (Wellcome Trust 2006b).

To control the risks of genome research, scientists, medical doctors, and funding bodies should promote a dialogue with citizens to make them familiar with the uncertainties of science, rather than with exaggerated promises. However, it has been reported that the general public and even researchers, medical doctors, and policy-makers have a difficult time understanding the hypothetical nature of science. Moreover, we still have very few clues about the use of genome information-based research and biomedical research and the resulting medical practices.

Observations of the Japan and UK biobanks made me aware of the dynamism of dialogue among stakeholders. A philosophy of monitoring and compliance will never bring about an era of amenable human experimentation in research, since the system is essentially based on outside ruling that contrasts with the cultivation of voluntariness of research domain and public. If the system could build on benefit-exchange of the research and the participants, 'volunatriness' would not be necessary. However, it seems not easy to understand indirect and/or remote exchange of benefit. As discussed, medical research involved a huge amount of indirect connection or exchange among generations involving the research participants, benefactors, and the future patients and beneficiaries. Such a philosophy based on remote benefit will not motivate the public to participate in important medical research. And the trend to demand direct benefits of science and medical research will weaken and damage the trust in medical research, because most results of 'advanced' research, including genome research to date, have not been benefiting the present patients in practice. At this moment we have to 'believe' that we should cultivate the cooperation of motivated stakeholders and stimulate dialogue, and then we may be able to construct a system of supporting medical research. This process in itself could realise the best use of genome research based on the human material and information of individuals, such as biobanks can support. As a beginning, we need to strengthen our role and position in the indirect and remote processes involved in genome and medical research.

Acknowledgements

There are so many people who helped me understand and think about the present situation and history of biobanks; in particular, I would like to thank Dr Koichi Bai, Professor Shin Utsugi, Dr Frances Rawle, Dr Ron Zimmern, and Dr William Lowrane for their inspiring discussion and encouragement during the course of this study. I am also very grateful to my colleagues in JCRB Cellbank, National Institute of Biomedical Innovation, especially, Dr Hiroshi Mizusawa for his generous support and Ms Yoko Takada for her constructive cooperation. This study has been supported by funds of MHLW (2001–2006), MEXT (2003–2004), and Human Science Foundation (1998–2003). Last, I would like to express my appreciation to Dr Margaret Sleeboom-Faulkner for giving me a chance to write this chapter and being patient during the course of its development.

Notes

1 July 2002 at the JMA.
2 Available at: http://www.jpma.or.jp/psc/frame-j.html, in Japanese. http://www.jpma.or.jp/psc/frame-e.html, in English, last visited on 21 November 2006.
3 Available at: http://www.biobankjp.org/index.html, last visited on 21 November 2006.
4 Riken is a governmental institution of comprehensive research on science and technology. http://www.riken.jp/engn/r-world/riken_menu/index.html, last visited on 21 November 2006.
5 Available at: http://www.jhsf.or.jp/index_b.html, last visited on 21 November 2006.
6 Available at: http://cellbank.nibio.go.jp/cellbank.html, last visited on 21 November 2006.
7 The Subcommittee on Science Project Evaluation and Life Science, Genome Research Working Party (*Genomu Kenkyu Ryouiki Shou-Iinkai*) was chaired by Dr Yoshiyuki Sakaki. It is located under the Committee of Science and Technology, and Academia (*Kagaku Gijutsu Gakujutsu Shingikai*) of the MEXT.
8 For a thorough review of the regulatory frameworks of Japanese and British genome research, see Porter, P. (2004) 'The regulation of human genetic databases in Japan', SCRIPT-ed http://www.law.ed.ac.uk/ahrb/script-ed/issue3/japan.pdf). He made an extensive comparison between the Japanese guidelines (2001 version) and the Ethics and Governance Framework of UK Biobank. The Japanese genome research guidelines have been revised, though the basic framework has not changed.
9 Available at: http://www.ukbiobank.ac.uk/, last visited on 21 November 2006.
10 Available at: http://www.advisorybodies.doh.gov.uk/piag/, last visited on 21 November 2006.
11 Available at: http://biobankjp.org/plan/elsi_com.html, last visited on 21 November 2006.
12 Available at: http://www.egcukbiobank.org.uk/, last visited on 21 November 2006.
13 Available at: http://www.hgc.gov.uk/Client/index.asp?ContentId=1, last visited on 21 November 2006.
14 Available at: http://www.dh.gov.uk/assetRoot/04/02/08/05/04020805.pdf, last visited on 21 November 2006.

15 Available at: http://www.opsi.gov.uk/acts/acts2004/20040030.htm, last visited on 21 November 2006.
16 Available at: http://www.hta.gov.uk/, last visited on 21 November 2006.
17 Notice from the Head of Medicine Agency, the Ministry of Health and Welfare, in Japanese: *Kouseishou Yakumukyoku-chou Tsuuchi*, Ihatsu-dai 1416Gou, 3 December 1970. Available at: http://www12.plala.or.jp/japa/tuken08.htm, last visited on 21 November 2006.
18 Report on the March 2002 meeting of the *Genomu Kenkyu Ryouiki Shou-Iinkai* (Genome Research Working Party), chaired by Dr Yoshiyuki Sakaki.
19 Available at: http://biobankjp.org/faq/faq_05.html, last visited on 21 November 2006.
20 I participated in a research programme supported by the MEXT, 'Social risk management of bioethical issues' in collaboration with Mitsubishi Research Institute http://www.chousei-seika.com/2005_s/2005_s_10/2005_s_10_1_seimeir inri/2005_s_10_1_seimeirinri_2.pdf [Japanese].
21 Available at: http://www8.cao.go.jp/cstp/kihonkeikaku/index3.html, in Japanese and English.

References

Annas, G. (2000) Rules for research on human genetic variation – lessons for Iceland, *New England Journal of Mediciine* 342: 1830–33.
CST. Hito genomu kenkyu ni kansuru kihongensoku (Fundamental Principle of Research on the Human Genome). 2000. http://www.mext.go.jp/a_menu/shinkou/ shisaku/gensoku.htm [Japanese]. http://www.mext.go.jp/a_menu/shinkou/shisaku/ principles.htm [English], last visited on 21 November 2006.
EGC. Meeting report of 'UK Biobank Ethics and Governance Council First Meeting.' http://www.egcukbiobank.org.uk/assets/wtd015345.pdf, last visited on 21 November 2006.
GeneWatchUK. http://www.genewatch.org/sub.shtml?als[cid]=507674, last visited on 21 November 2006.
Giles, J. (2006) Huge biobank project launches despite critics, *Nature* 440: 263.
Government Office. http://www.kantei.go.jp/jp/mille/index.html [in Japanese], last visited on 21 November 2006.
Hapgood, R., Shickle, D. and Kent, A. (2006) Consultation with primary care health professionals on issues relating to the recruitment of patients to a DNA collection study. http://www.ukbiobank.ac.uk/docs/GPreport.pdf, last visited on 21 November 2006.
HGC. Human Genetics Commission. Inside information: balancing interests in the use of personal genetic data. http://www.hgc.gov.uk/Client/document.asp?DocId=19& CAtegoryId=8, 2002, last visited on 21 November 2006.
House of Commons, Science and Technology Committee. The Work of the Medical Research Council. 2003. http://www.publications.parliament.uk/pa/cm200203/ cmselect/cmsctech/132/132.pdf, last visited on 21 November 2006.
Jonas, H. (1974) Philosophical reflections on experimenting with human subjects. Against the stream in *Philosophical Essays: From Ancient Creed to Technological Man*, Englewood Cliffs, NJ: Prentice-Hall, 105–40.
Kojin jyouhou hogo ni kansuru houritu (Act on the Protection of Personal Information) (30 May 2003) http://www5.cao.go.jp/seikatsu/kojin/houritsu/index.html [Japanese]. http://www5.cao.go.jp/seikatsu/kojin/foreign/act.pdf [English], last visited on 21 November 2006).

Masui, T. (2002a) Shigen to naru jintai (Human body as resource), *Gendai Shiso* (Modern Philosophy) [Japanese] 2: 194–210.

Masui, T. (2002b) Genomu kenkyu no jidai (The era of human genome research), *Sekai* (World) [Japanese] 11:199–208.

Masui, T. (2003) Eikoku baiobanku ga imisuru mono (What UK Biobank means), *Jurist* [Japanese] 1247: 29–36.

Masui, T. (2004a) Genomu kenkyu wo sasaeru shakai-kiban wo mezashite (On the infrastructure supporting genome research), *Japan J. Sci. Tech. Society* [Japanese] 13: 91–109.

Masui, T. (2004b) Chronicle of UK Biobank in the early phase including the wide range of supporting activities and documents. Available at: http://cellbank.nibio.go.jp/information/ethics/kiban01/image/chronicle20040130.jpg, last visited on 21 November 2006.

Masui, T. and Takada, Y. (2005) Eikoku baio-banku purojyekuto (The UK Biobank project), *J. Exp. Med.* [Japanese] 23: 522–29.

METI. http://www.meti.go.jp/policy/bio/kojinidenhogo-houkoku/sankou4.pdf [Japanese], last visited on 21 November 2006. This is a summary of misconducts in genome research.

MEXT. Raifu Saiensu niokeru Kenkyu Kaihatsu no Shinkou ni Tsuite (Promotion Strategy of Research and Development of Life Science). June 2002. http://www.mext.go.jp/b_menu/shingi/gijyutu/gijyutu2/toushin/020701.pdf [Japanese], last visited on 21 November 2006.

MHLW. Idenshi kenkyu ni fuzui suru rinrimondai tou ni taiousurutameno shishin (The Millennium Guidelines: Guidelines on ethical issues of genetic analysis), http://www1-bm.mhlw.go.jp/topics/idensi/tp0530-1_b_6.html [Japanese], last visited on 21 November 2006, presently not effective.

MHLW. Iryou/kaigo kankei jigyousha ni okeru kojinjyouhou no tekisetsu na toriatukai no tameno gaidorain (Guidelines for Appropriate Handling of Personal Data in Medical and Care Services) 2004. http://www.mhlw.go.jp/topics/bukyoku/seisaku/kojin/dl/170805-11a.pdf [Japanese] , last visited on 21 November 2006.

MHLW, MEXT, and METI. Hito genomu idenshikaiseki kenkyu ni kansuru rinrishishin (The Ethical Guidelines for Analytical Research on Human Genome/Genes) 2001. http://www.mext.go.jp/a_menu/shinkou/seimei/genomeshishin/html/rinri_shishin.htm [Japanese]. http://www.mext.go.jp/a_menu/shinkou/seimei/genomeshishin/html/shishin/rinrishishin_english.pdf, [English], last visited on 21 November 2006. Revised in 2004, http://www.mhlw.go.jp/general/seido/kousei/i-kenkyu/genome/0504sisin.html [Japanese].

MRC. Human Tissue and Biological Samples for Use in Research – Operational and Ethical guidelines. 2001. http://www.mrc.ac.uk/Utilities/Documentrecord/index.htm?d=MRC002420, last visited on 21 November 2006.

NBAC. National Bioethics Advisory Commission. Ethical and policy issues in research involving human participants, 2001. http://www.georgetown.edu/research/nrcbl/nbac/pubs.html, last visited on 21 November 2006.

NCBE. Human tissue: Ethical and legal issues. 1995. http://www.nuffieldbioethics.org/go/ourwork/humantissue/publication_298.html, last visited on 21 November 2006.

Nukaga Y. and Tsutani K. (2006) Idenshireigaishugi (Genetic exceptionalism). *Nihon Igakukai Zasshi* (Journal of the Japan Medical Association, JJMA) [Japanese] 134:2385–90.

OECD Guidelines on the Protection of Privacy and Transborder Flows of Personal

Data. 1980. http://www.oecd.org/document/18/0,2340,en_2649_201185_1815186_1_1_1_1,00.html, last visited on 21 November 2006.

OECD. Creation and Governance of Human Genetic Research Databases. 2006. OECD Publishing. e-Book available at http://www.oecd.org/document/50/0,2340,en_2649_34537_37646258_1_1_1_1,00.html, last visited on 21 November 2006.

Ohnishi, Y. and Nakamura, Y. (2005) Baiobanku jyapan purojekuto (Biobank Japan Project), *Nippon Rinsho* [Japanese] 63 Suppl 12:35–41.

PSP. People Science & Policy Ltd. (2002) BioBank UK: A question of trust: a consultation exploring and addressing questions of public trust. http://www.ukbiobank.ac.uk/docs/consultation.pdf, last visited on 21 November 2006.

Rawle, F. (2003) UK DNA sample collections for research. In: B.M. Knoppers (Ed.) *Population and Genetics*, Herndon, VA: Martinus Nijhoff Publishers, 3–15.

Rose, H. (2006) From hype to mothballs in four years: troubles in the development of large-scale DNA biobanks in Europe. *Community Gentics* 9: 184–9.

Shamos, M. (1995) *The Myth of Scientific Literacy*, Chapter 3, The nature of science, New Brunswick, NJ, Rutgers University Press.

Shryock, R. (1974 [1936]) *The Development of Modern Medicine*. Madison, WI: University of Wisconsin Press.

UK Biobank. Sample handling and the storage, subgroup protocol and recommendations, version 1.0. 2004. http://www.ukbiobank.ac.uk/docs/Sample_Handling_and_Storage_Subgroup_report_for_comment.pdf, last visited on 21 November 2006.

UK Biobank. UK Biobank Ethics and Governance Framework. Version 1.0, 24 September 2003. http://www.ukbiobank.ac.uk/docs/egf-comment-version.doc, last visited on 21 November.

Umeda, M., Sato, K., Shibanuma, M., Tanaka, N., Hata, H., Hirai, R., Masui, T., Utsugi, S. and Matsumura, T. (1998) Hi-rinshou ni okeru hito-soshiki/saibou no toria tukai ni tsuite (On the use of human tissues and cells in non-therapeutic research and development – discussion paper on research ethics and safety), *Tissue Culture Res. Commun.* [Japanese] 17: 117–71. http://jtca.dokkyomed.ac.jp/JTCA/ethics/9812jtca.htm, last visited on 21 November 2006.

UNESCO. The Universal Declaration on the Human Genome and Human Rights. 1997. http://portal.unesco.org/en/ev.php-URL_ID=13177&URL_DO=DO_TOPIC&URL_SECTION=201.html, last visited on 21 November 2006.

Wellcome Trust. UK Biobank. http://www.wellcome.ac.uk/doc%5Fwtd003504.html, last visited on 21 November 2006a.

Wellcome Trust (text: Jon Turney and Sarah Davies). Engaging Science Conference. Science for all: is public engagement engaging the public. http://www.wellcome.ac.uk/doc_WTX032160.html, last visited on 21 November 2006b.

WMA. World Medical Association Declaration of Helsinki: Ethical Principles for Medical Research Involving Human Subjects, the 52nd WMA General Assembly, Edinburgh, Scotland, October 2000. Available at: http://www.wma.net/e/policy/b3.htm, last visited on 21 November 2006.

5 Should we invest in biobanking in Hong Kong?

Using biobanking for dyslexic studies in Hong Kong as an example

Mary Miu Yee Waye and Connie Ho

Introduction

The aim of this chapter is to help decision-makers think clearly about the complex decision of whether to invest in biobanking in Hong Kong, by discussing some of the benefits and challenges of the investment. To limit the scope of the discussion, the pros and cons of biobanking in Hong Kong for advancing the study of dyslexia in Hong Kong will be chosen as an example to illustrate how priorities could be set. Some of the issues of trust and privacy of such an effort will be discussed.

The contribution of biobanking to advancement in medical research

There have been numerous studies that have demonstrated the usefulness of biobanking in gene–environmental interactions and subpopulation susceptibility to diseases. For example, several regions on the chromosomes were found to be a possible susceptibility region for psoriasis (Karason *et al.* 2005). By comparing Finnish male breast cancer patients and those in Iceland, it was suggested that modifying genetic and environmental factors may significantly influence the penetrance (defined as the percentage of individuals with a given genotype who exhibit the phenotype associated with that genotype) of cancer genes in individuals carrying germline mutations in the BRCA2 gene (Syrjakoski *et al.* 2004). In Hong Kong, an extensive study was done on the association of different genetic and environmental factors leading to susceptibility to obesity and type II diabetes (Li *et al.* 2006) and it was suggested that the screening of metabolic syndrome was important for the management of diabetes, especially when the family has early-onset diabetes. Research has also been done using a smaller bank of SARS patients to detect polymorphism that might affect susceptibility to SARS coronavirus infection (Mak *et al.* 2006).

Many researchers have pointed out the urgent need for biobanking activities to further validate their findings. For example, a study by Senn *et al.* has demonstrated findings of a possible interaction of severe and intermediate AAT deficiency (caused by deficiency of the protein alpha-1-antitrypsin or AAT, which leaves the lung, and occasionally the liver, vulnerable to injury) with modifiable chronic obstructive pulmonary disease (COPD) risk factors (such as gas, fumes and dust, or to indoor air pollution) other than active smoking that needs to be further examined with a larger population (Senn *et al.* 2005).

This chapter advises on whether similar projects should be carried out in Hong Kong, using the putative usage of such a biobank in the study of dyslexia in Hong Kong as an example.

Dyslexia and genes

It has been estimated that between 5 per cent and 8 per cent of the total population may be dyslexic, though only approximately one out of a hundred people have been identified as dyslexic and offered assistance. Though compensation strategies for learning can be developed by dyslexic children, often have great difficulties in traditional classrooms and many of them will form a major part of the category of adults with the lowest level of literacy. Most parents have grave concerns for their dyslexic children, though some consider that dyslexia is a gift because dyslexic people are visual, multidimensional thinkers. Often, they are intuitive and highly creative, and excel at hands-on learning. The names of some of the many individuals who are dyslexic or have symptoms of dyslexia (e.g. Albert Einstein and Tom Cruise) have been posted on this website: http://www.dyslexia.com/faq.htm. However, it was also reported that adults with dyslexia continue to have communication problems that are barriers for employment and life issues in general. It has been reported that the frequency of dyslexic problems among prison inmates is higher than in the normal population (Samuelsson *et al.* 2000; Samuelsson and Herkner 2003; Svensson *et al.* 2000). Thus, research into dyslexic genes needs to take into consideration the volunteers' concern of discrimination by future employers.

Dyslexic Hong Kong children

Based on the data of the normative study of the Hong Kong Test of Specific Learning Difficulties in Reading and Writing, Chan *et al.* (in press) reported a prevalence rate of 9.7–12.6 per cent of developmental dyslexia in Hong Kong and a boy:girl gender ratio of 1.6 to 1. Hong Kong dyslexic children have problems learning English as a second language as well (Ho and Fong 2005). Perhaps, because of a poor public awareness and understanding of dyslexia in the Hong Kong community and possible referral bias in clinical samples, the incidence rate for cases referred to the local education authority was 0.32 per cent and the boy:girl gender ratio was 3 to 1 (Chan *et al.* under review). Research findings show that the main difficulty of alphabetic dyslexia readers, especially English-speaking ones, lies in phonological processing, whereas for non-alphabetic scripts, e.g. Chinese, orthographic-related difficulties may be the crux of the problem. It was found that the majority of dyslexic children in Hong Kong have rapid naming and orthographic deficits (accounting for 50 per cent and 39 per cent, respectively), and a relatively small proportion have phonological deficits (20 per cent) (Chan *et al.* 2004; Ho *et al.* 2004). Hong Kong children learn Chinese in a manner different from the mainlanders and Taiwanese in that the phonetic system or 'pin-yin' method is not generally taught to aid with learning Chinese in Hong Kong at an early stage, but in general the 'look-say' method is used for Hong Kong children to learn Chinese scripts instead. Children in Hong Kong have to retrieve the pronunciation of Chinese characters by rote rather than through assembling the sound of pin-yin letters as in Mainland China. This unique characteristic mode of learning is consistent with finding certain types of deficits to be more prominent in dyslexic Chinese children in Hong Kong. Though there have been reports of the discovery of genetic loci/genes responsible for susceptibility of dyslexia (http://www.ncbi.nlm.nih.gov/entrez/query.fcgi?CMD=search&DB=pubmed), and DYX1C1 was the first gene to be described as a candidate for dyslexia (Grigorenko 2003, 2005; Grigorenko *et al.* 2003), it is logical to predict that the genes that might predispose children to dyslexia will be different for children in Hong Kong, compared to the rest of the world, including Mainland China, Taiwan, and Singapore (which currently adopts the Mainland China way of teaching Chinese characters). Thus, we argue that dyslexia is an example of a problem unique for Hong Kong and a biobank with a collection of materials from Hong Kong would be helpful in further studies of the genes involved with dyslexia in Hong Kong.

Hong Kong, a Westernised city that embraces scientific discoveries and technological advances, is also suffused with traditional Chinese culture. It is the general attitude that children should be subjected to strong pressure for academic achievement and there is very little alternative for those children who do not conform to the norm. This causes an extra degree of pressure for those families with dyslexic children who have to constantly defend themselves against prejudices of the teachers and classmates, and those who might not understand the innate disability of the dyslexic children. Thus, understanding the genetic basis of susceptibility to dyslexia might have a special meaning for these families and could lead to a certain degree to lowering the anxiety of the parents, who might then concentrate on how to teach the dyslexic child effectively with different types of simulations, rather than merely pressurising a child to achieve academically.

Issues related to privacy

In Hong Kong, because relatively few genetic studies have been conducted, especially on behavioural or learning problems, parents of dyslexic children may not be fully aware of the possible consequences of their genetic data being collected and analysed. It will be important, therefore, that individual informed consent with clear explanation of rights, benefits, and risks be carried out in genetic studies, particularly when samples collected could be used and stored for a long time, such as the case with biobanking.

Many of the parents of dyslexic children in Hong Kong also worry about the labelling effect, i.e. they carry the label of having a genetically linked disability. Hong Kong parents do not normally want to disclose their child's learning disability to the school. They are afraid that this will create a negative image for the teachers and affect their child's learning opportunities. For some educators, a genetically linked problem may mean difficulty of remediation, though this may not be the case. The issue of possible discrimination in school will reduce the willingness of participants to donate to the biobanks.

There are strategies that may encourage people to participate in these genetic studies. Since there is general lack of understanding about the relationship between the genetic contribution to a learning disability and the impact on the remediation outcome, both parents and children may be given counselling when they learn about their genetic status of the disability, and individual results may be explained by the investigators. Alternatively, investigators may analyse the data collectively and inform the participants about group results. As a general practice, we believe that, at all times, only group results should be reported to the public.

Since the concept of biobanking studies is relatively new in Hong Kong, it would be premature to link a genetic database with a database of healthcare or education systems in Hong Kong. More research findings are required on the pattern of heritability and the association of, and interaction between, geno-types and phenotypes (e.g. nature and scope of the problem, manifestation, and persistence of the problem) before we are convinced that disclosing the genetic information to the healthcare and education systems would benefit the clients.

To ensure good participation in genetic research among the Chinese, it is especially important to develop trust. The Chinese are a collectivist society in which people place a lot of emphasis on 'face'. Having a genetic disability may bring disgrace to the family. Some parents are reluctant to learn about the causes of their children's learning problems. They are afraid to find out that they may be partly responsible for their children's difficulties. The possible heritage of dyslexia is not widely known or discussed in the local community. Thus, parents might be reluctant to participate, and need more reassurance, education, and trust.

In addition, the general sense of trust within the local society of Hong Kong is relatively low. Extra measures would need to be taken to protect the privacy and benefits of the clients in order to improve their sense of trust in genetic research. Such measures may include not transferring the data to third parties without the consent of the clients. A healthy ethics review and monitoring system is also essential to develop trust between investigators and clients. Ethics review systems are generally good in local universities and healthcare systems. The clients should be informed of the proper committees or units to which they may channel their complaints, if any.

Another way to encourage participation in donating to the biobank is for the genetic information/samples to be jointly owned by the clients and the investi-gators. This means that the clients have the right to withdraw their information/

samples at any time. The investigators can only use the information for research purposes. Utilisation of the data other than for the original purpose should require consent from the clients.

We conducted a pilot survey during the summer of 2005 to find out the feeling of Hong Kong people about the use of biobanking for the studying of a candidate gene for dyslexia. Due to limitation of available funds, we chose the authoritative sampling method with the help of four summer students who had been briefed about the objectives of the study. The survey, not included in this chapter, was conducted in Chinese, and the corresponding English translation is shown in the appendix.

Our survey on biobanking is divided into four parts. The first part gives an overall idea of the participant's personal information, which would be useful for understanding the characteristics (e.g. age and educational levels) of the participants who are for and against biobanking activities. It also provides information for future comparison with other surveys. The second part is related to the knowledge of the participants on biobanking activities. The third part is a survey on the participant's feeling about the confidentiality issue which should give some ideas for future studies in public policy related to DNA biobanking. The remaining part deals with ethical issues and the issue of trust. The willingness of the participants to donate samples, as well as their concerns about participating in biobanking activities, is also surveyed.

The participants of this survey were mainly university students under 21 with the second largest group between 21 and 30. Most of them were born in Hong Kong while some are from Mainland China. Most of the participants reported little knowledge of genetic diseases, but they realised that genetic data is a type of personal and confidential data and felt that such data should not be used for other research groups who are collaborating with the original research group without the consent of the donor (presumably the research group had already obtained informed consent from the donors earlier). The participants had mostly heard about biobanking (with the more educated group being the most likely to have heard about it) and they were generally supportive of establishing a biobank in Hong Kong. However, their willingness to donate DNA samples was a bit less than their expressed support (half being unwilling to do so even when they were assured that the researchers and the donors would both own the DNA samples). The participants were slightly keener if there was some type of reward for donating DNA, but still there was a group who was not willing despite the enticement. Many participants felt that shame was not associated with having genetic diseases but they felt that they would be responsible if their child had a genetic disease. Many participants were not willing to discuss the genetic disease if there were an affected member in their family, but the more educated group was the most willing to discuss it. Many participants in our survey viewed dyslexic children as sick or lazy, and few were optimistic about the usefulness of understanding the genetic basis of dyslexia. Such beliefs could be due to a lack of understanding of the reading disorder and recent advancement in genetic diseases; acceptance of fate was seen as a better alternative than expensive and ineffective treatment or a fear of what other possible consequence (possibly worse) might be caused by the lengthy path towards the pursuit of knowledge. The fact is, dyslexia occurs at all levels of intelligence and there are examples of the usefulness of understanding the genetic basis of diseases.

This preliminary survey was designed to provide some insight into some of the issues of trust and privacy related to DNA biobanking. Some aspects of the findings are not surprising, including the generally low level of knowledge of DNA biobanking in Hong Kong and the general public's reasonable level of enthusiasm for biobanking, albeit with a lesser degree of actual willingness to donate DNA. This is a reflection of the participants' feelings about the negative effect related to social life that DNA biobanking might impose on them (e. g. discrimination by peers) (Hoeyer 2005), and implies that there are moral opinions held by the participants that make them less willing to donate their own DNA samples.

There were some unexpected findings in this survey. For example, older subjects responded differently in their willingness to participate in biobanking activities, including their lower level of self-reported knowledge of DNA biobanking. They were less likely to regard DNA as confidential data, they were less likely to donate DNA, and more likely to feel that carrying a genetic disease is something to feel ashamed about. One could argue that being older, such people have had more time to assimilate complicated knowledge and to appreciate the potential benefit that genomics research using DNA biobanks might bestow on them. However, it is also possible that being older means a decrease in openness towards change. This would explain their low level of willingness to donate and share their genetic material. (See 'Personality, stress and coping section' by Robert McCrae, in http://www.grc.nia.nih.gov/branches/lpc/rmccrae.htm, updated 12 May 2005). It was also found that with age, occurrence of personality traits such as neuroticism, and extraversion decrease, whereas the occurrence of traits such as agreeableness and conscientiousness increase.

The feeling of helplessness reported by some people with dyslexia (Boggiano and Barret 1985) is understandable, as geneticists do not provide therapy for many genetic conditions that have known disease genes. For this reason it seems reasonable to assume that the efforts of scientists in discovering the particular gene concerned would be more an academic exercise than yield any practical benefit. However, such pessimism does not take into consideration the benefit of early diagnosis for many genetic conditions. Credit should be given to geneticists (Woo *et al.* 1982, 1983) who found out how to diagnose early phenylketonuria (PKU), a condition that can lead to mental retardation if the patient is not given a proper diet at an early stage. Some advancement in gene therapy for phenylketonuria has been discovered recently (Chen and Woo 2005; Cristiano *et al.* 1993). Another successful application of knowledge obtained from genetic information is the discovery of the molecular basis of glucose-6-phosphate dehydrogenase deficiency, which facilitated understanding of the differential drug sensitivity of certain individuals who now can avoid anaemia through drug administration, and prevent stress due to infection (Vulliamy *et al.* 1992, 1993). Another example of the application of genetic information is that of thalassemia; with the advent of early diagnosis, many cases of would-be sufferers needing life-long blood transfusion can be avoided while milder cases can be recognised early and properly cared for (Kan *et al.* 1980; Kazazian *et al.* 1992; Taylor *et al.* 1974). The general public may feel that the brain is such a complex organ that we have little hope of understanding its normal operation, let alone the unwieldy networking that might lead to dyslexia. But, again, we could draw on the success of the search for the Huntington disease gene. For many years, the Huntington gene remained elusive, due to the novel trinucleotide repeats that constitute a novel type of mutation different from previous single-gene diseases (Antequera and Bird 1993; Goldberg *et al.* 1996).

Perhaps one could argue that dyslexia is not a condition of profound and immediate clinical ramifications. For instance, it could hardly be expected that the link between carrier and genetic make-up is as 'straightforward' as in the case of familial susceptibility to breast cancer: carriers of the BRCA1 gene have a life-time risk of 85 per cent of developing breast cancer and an unknown risk of ovarian cancer (Biesecker *et al.* 1993). The general public therefore may not think that the discovery of a dyslexia gene would be of much importance. However, one need only have a short conversation with the parents of such a child to realise the tremendous social consequences it has for both the child's well being and the family's struggle with learning disability.

It may be considered unethical to offer the information on the carrier status for dyslexia (if the disease gene were to be known) to the family or to the extended family members since the general public could misunderstand the meaning of such a status and treat the carrier as a defective person with less value. A question that would arise is who is to decide whether a child should be tested, the parent or the child. Furthermore, an unpredictable side effect of the testing may be its impact on intrafamilial relationships (Wertz *et al.* 1994). For instance, the parents may feel helpless and provide less care or opportunities for one they may label as a 'defective'. One could argue that it is not easy to diagnose dyslexia at an early stage in life, given that environmental factors could affect the child's ability to learn. However, there are studies that indicated some physiological markers can distinguish dyslexic children as early as the neonatal stage (Molfese 2000), which is a logical extension from earlier work showing that there are functional disruption patterns in brain systems of dyslexic subjects during reading tasks (Shaywitz *et al.* 2002). On a positive note, it has been reported in some studies that the neurological abnormalities as shown by brain activation profiles resumed a normal pattern following successful remedial training (Simos *et al.* 2002).

We do not yet know the biochemical pathways that might lead to dyslexia but several mechanisms have been proposed (Binder *et al.* 1996; Rumsey *et al.*

1997), including the cerebellar deficit hypothesis which states that the cerebellar function is abnormal for dyslexic children (Nicolson and Fawcett 2005).

Scientists have found that there are three areas of the brain involved in reading and that dyslexic children do not activate the back of the brain as do other readers. A lack of oxygen at the time of birth could lead to dyslexia but some believe that this probably accounts for only a very small percentage of cases (see http://www.readersdigest.ca/mag/2004/12/dyslexia.html), and by and large it seems to be genetic.

Thus, it is possible that genetic–environmental interaction plays an important role in addition to genes alone. Moreover, a therapeutic option could be developed if the candidate gene is discovered.

Conclusions

We have identified some of the issues of trust and privacy of biobanking for the research of dyslexia in Hong Kong. Without the help of the biobank, research on susceptibility genes for dyslexia could still be done. However, as studies on schizophrenia and other neurological diseases have shown, complex disorders such as these are not likely to be solved by single laboratories with the technology at hand. Their separate efforts would mean that multiple laboratories would each be collecting a population of dyslexic subjects and their associated phonological data, i.e. a duplication of effort and inefficiency of the research. In the light of these considerations, biobanking for future studies for finding genes for dyslexia would be a worthwhile effort for Hong Kong. Our survey uncovered the uneasiness of some subjects towards donating their DNA samples to biobanks (and their concern about sharing genetic materials for collaborating groups which might be located overseas). In spite of this, our survey provides some cause for optimism since a large proportion of respondents showed an interest in supporting the establishment of a biobank in Hong Kong. In order to facilitate collaboration between the general public and the scientists, it is crucial that the both parties have a clear understanding of the limitations and advantages of biobanking activities.

Acknowledgements

The authors wish to thank the Croucher Foundation for financial support. The views of the authors do not necessarily reflect the view of the Croucher Foundation. The authors would also like to thank: Ho Sai Fun, Ng Lung Yan, Andy, Cheng Wai Hang, Tom, Kong Kit Fan, and Frankey for their help with the design of the questionnaire, the survey, and the analyses of data.

Appendix

Survey of the attitude towards biobanking activities

Aim

To study the attitude towards biobanking activities in Hong Kong.

Method

Due to limitation of available funds, we chose the authoritative sampling method with the help of four summer student helpers who had been briefed about the objectives of the study. The survey was conducted in Chinese.

Results and discussions

Part 1: Personal information

1   The majority of the population sampled in our survey is under 21 (202 out of 301; due to the sampling design, the majority of the people given the questionnaires are university students and their friends and relatives. The second largest group are those between 21 and 30 (63 out of 301).
2   The majority of the subjects are born in Hong Kong (80 per cent) whereas the second largest group are from China (20 per cent).
3   The level of education of the subjects is mostly college level or above which reflects the fact of the survey design also.
4   Nearly two-thirds of the subjects do not have any religious affiliation, with the second largest group being Christians.
5   The majority of the subjects are unemployed, which reflects the fact that they are mostly university students and thus unemployed during the summer break.
6   The gender of the subjects is evenly distributed between male and female.

Part 2: Subjects' knowledge about genetic diseases

7   What is your level of knowledge on genetic diseases?
The level of knowledge on genetic diseases based on the subjects' own assessment was low (Figure 5.1). The older (or less educated) are less knowledgeable about genetic diseases.

Part 3: Confidentiality issues

8   Do you agree that genetic data is a type of personal confidential data?
Most subjects agree that genetic data is a type of personal, confidential data (Figure 5.2), and the older (or less educated) were more likely to agree that genetic information is a type of personal confidential data.

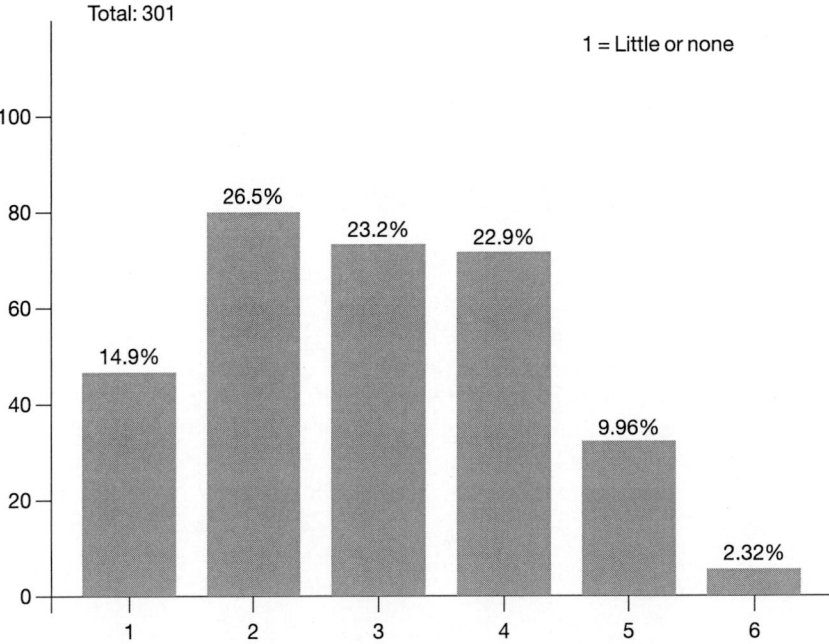

Figure 5.1 The degree of self-reported understanding of genetic diseases.

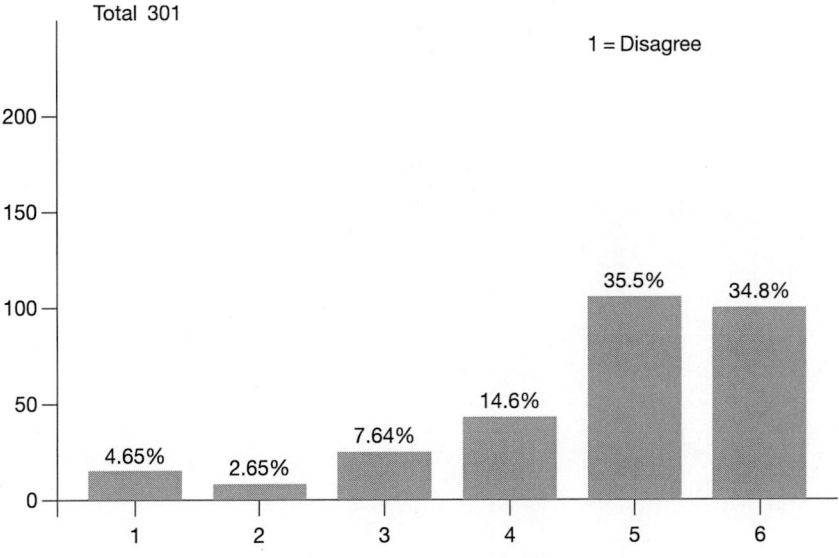

Figure 5.2 'Genetic information should be kept as confidential data'.

9 Assuming everything is done anonymously, can you agree that genetic material can be used for other research projects without consent of the donors? While the subjects felt that genetic material should not be used for other research projects without consent of the donors (Figure 5.3), the more educated (but not necessarily younger) subjects were more likely to think that it is acceptable that, if done anonymously, even in the absence of consent, the genetic material could be used for a different type of research other than the one specified.

Part 4: Ethical issues

10 Have you heard of DNA biobanking?
Most subjects had heard of DNA biobanking (Figure 5.4) and the more educated (not necessarily younger) group of subjects were more likely to have heard about DNA biobanking.
11 Do you support having a DNA biobanks in Hong Kong?
Most subjects were supportive of the idea of having a DNA biobank in Hong Kong (Figure 5.5). The more educated (not necessarily younger) group were more likely to support biobanking in Hong Kong. This might reflect the adventurous nature of the Hong Kong population. Though the

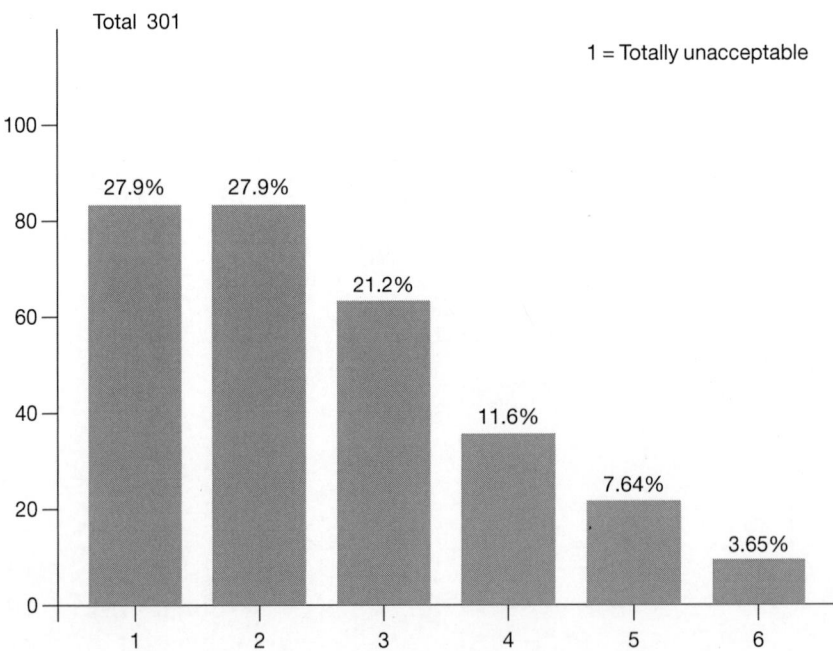

Figure 5.3 Should one use donors' genetic materials for other research projects without further consent?

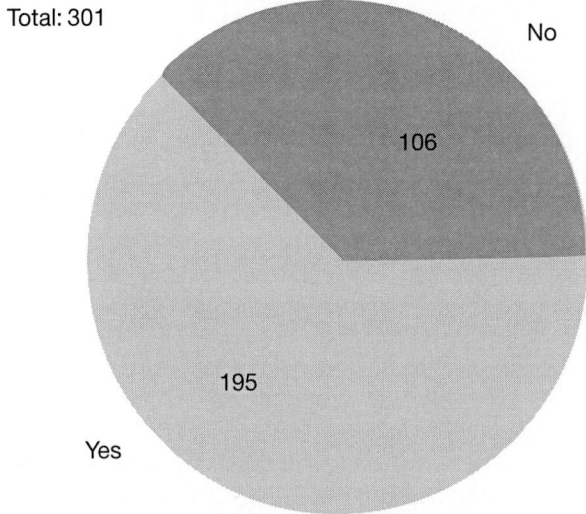

Figure 5.4 The number of people who had previously heard of biobanking.

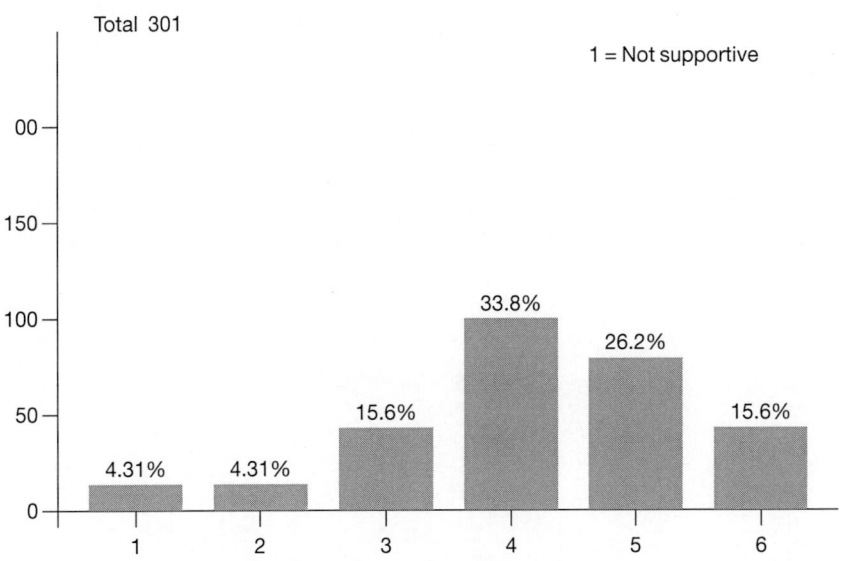

Figure 5.5 Do you support having a DNA biobank in Hong Kong?

majority of the population is Chinese in origin, the acceptance of Western culture and anything that could be perceived to improve the health and welfare of the people are welcome.

12 If Hong Kong were to carry out research related to DNA biobanking, would you be willing to donate your DNA?

Though the response could be considered as neutral, with more or less equal numbers of willing versus unwilling subjects, the subjects' enthusiasm for donating their DNA was not directly correlated with the enthusiasm of the previous question on whether they supported setting up a DNA biobank (Figure 5.6). This might be due to the feeling of uncertainly in the process, lack of knowledge of the significance, and wariness of the possible perils of submitting something confidential. The more educated (not necessarily younger) group were more likely to participate in genetic research if Hong Kong were to set up a DNA biobank.

13 If the DNA were to be owned jointly by the donors and the researchers, would you be willing to donate your DNA?

For those unwilling to donate, the incentive remains relatively unattractive even if the DNA were to be owned jointly by the donors and the researchers, though some could be persuaded to donate if DNA were to be owned jointly (Figure 5.7). The older (or less educated) were less likely to donate even if it is agreed that the material would be owned jointly by the donors and the researchers.

14 If there were some type of reward for donating DNA, would you be willing to donate your DNA?

More people would be willing to donate if some types of reward were to be given for donating DNA (Figure 5.8); however, the portion that is

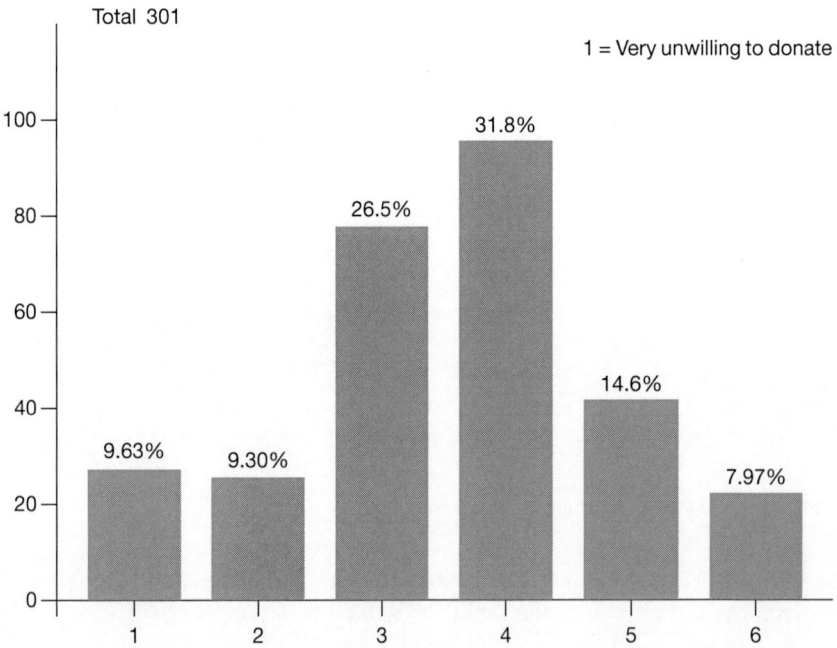

Figure 5.6   If Hong Kong were to carry out research related to DNA biobanking, would you be willing to donate your DNA?

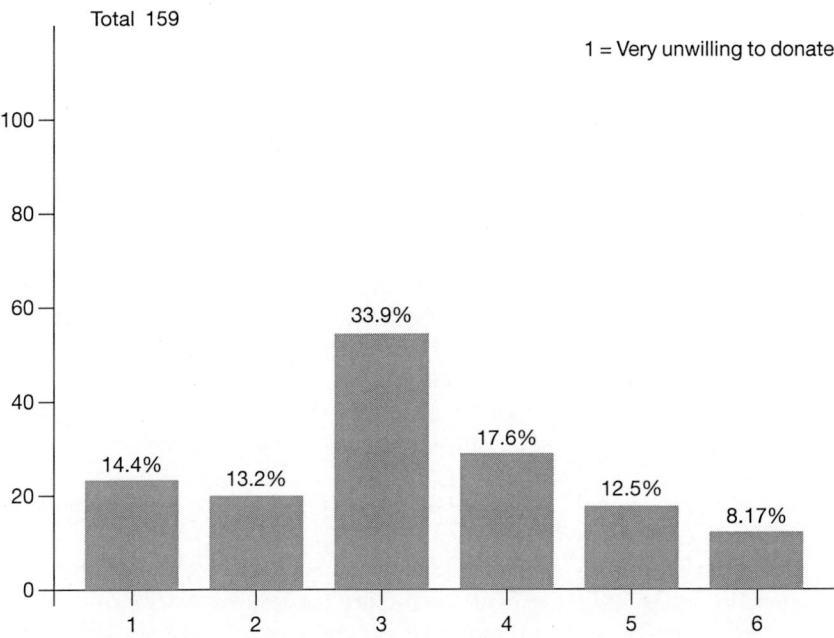

Figure 5.7 If the DNA were to be owned jointly by the donors and the researcher, would you be willing to donate your DNA?

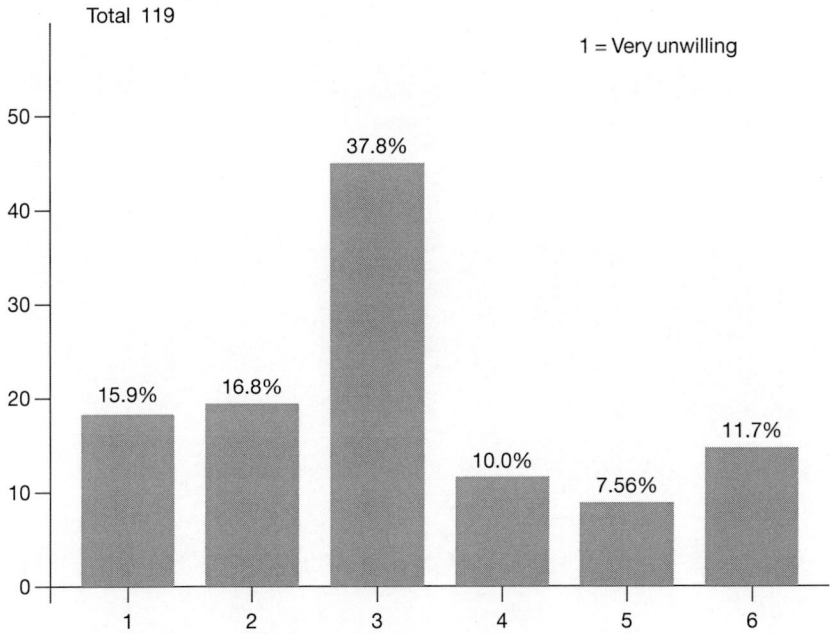

Figure 5.8 If there were some type of reward for donating DNA, would you be willing to donate your DNA?

unwilling still dominates. The older (or less educated) were less willing to donate even if some types of reward were to be given.

Part 5: Issue of trust

15 During the process of the research, you would be willing to let the DNA sample that you donate be used by other agency or educational institutions participating on a collaborative basis?
Most subjects were not willing to let the DNA be used by another agency or educational institutions participating on a collaborative basis (Figure 5.9). The willingness to have their DNA samples shared does not appear to be related to age or education level.

Part 6: Mandatory questions. Do you agree with the following?

16 Do you agree with the statement that having genetic diseases is something of which to be ashamed?
Most subjects disagreed with the statement that shame was associated with genetic diseases (Figure 5.10). The older (or less educated) group were somewhat more likely to think that having a genetic disease was something of which to be ashamed.

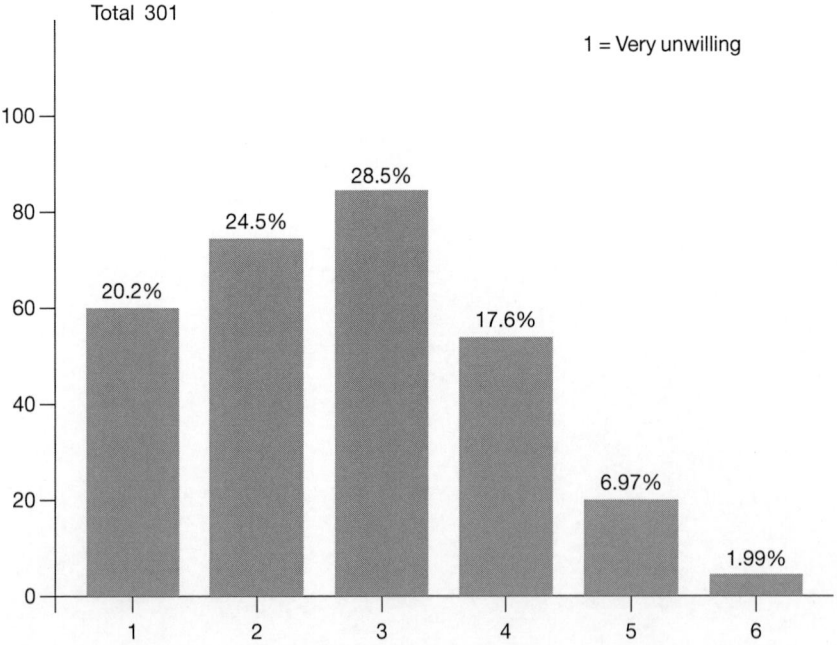

Figure 5.9 During the process of the research, would you be willing to let the DNA sample that you donate be used by other agencies or educational institutions participating on a collaborative basis?

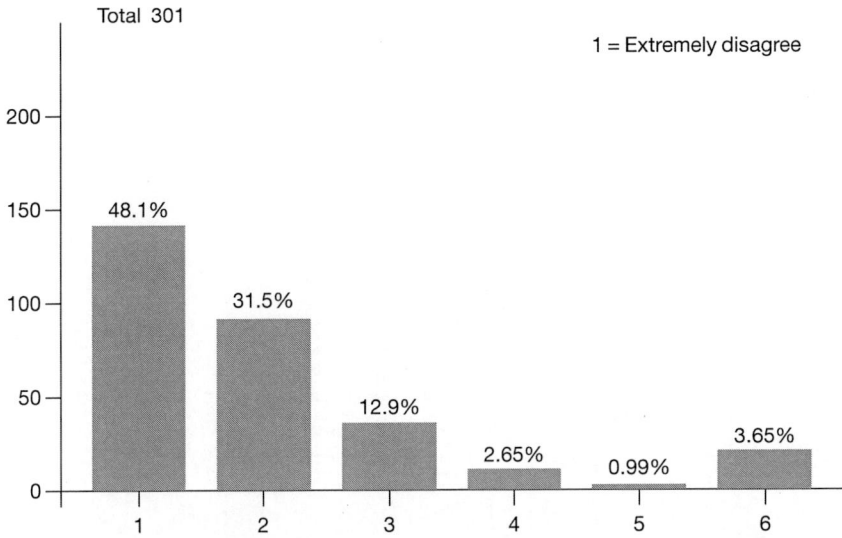

Figure 5.10 Do you agree with the statement that having genetic diseases is something of which to be ashamed?

17 Would you discuss genetic diseases with your friends?
   There was an even distribution of willingness to discuss genetic diseases with friends (Figure 5.11). The more educated subjects (not necessarily younger) are more likely to discuss genetic diseases with others.
18 If one of your family members had a genetic disease, would you actively find out its cause?
   There seems to be a lack of curiosity for most participants because they chose that they would not actively find out the cause of a genetic disease if their family members have it (Figure 5.12). The education level or age of the subjects did not appear to affect the response to this question.
19 If your child had a genetic disease, would you feel responsible for it.
   Most subjects felt that they should be responsible for it if their child had a genetic disease (Figure 5.13; 72 per cent responded positively) and the more educated felt slightly more responsible for it.

Part 7: Reference questions

20 If a child has a problem reading and writing, what do you think might be the reason for the problem?
   Most think that if a child has a problem reading and writing, the child is sick (62 per cent), many think that they might be lazy (20 per cent), some feel that the school is the problem (13 per cent), some feel that the parents are too relaxed (9 per cent), and a small proportion fell that it is living conditions that causes the problem (7.5 per cent) (Figure 5.14).

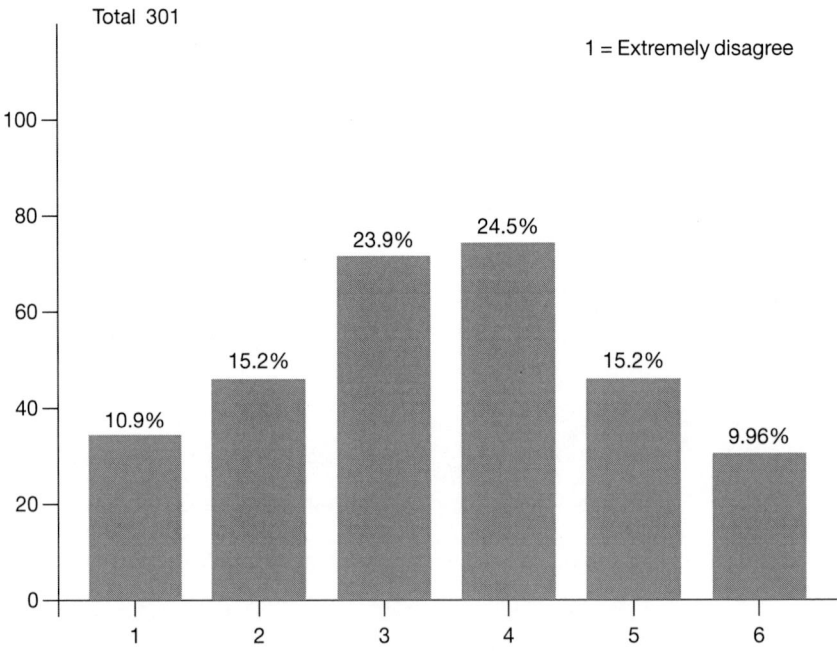

Figure 5.11 Would you discuss genetic diseases with your friends?

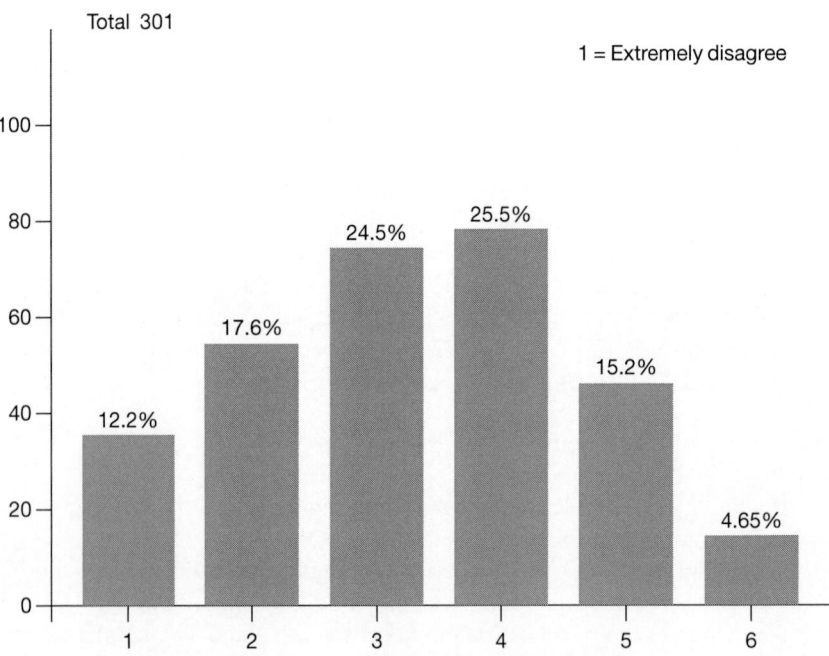

Figure 5.12 If one of your family members had a genetic disease, would you actively find out its cause?

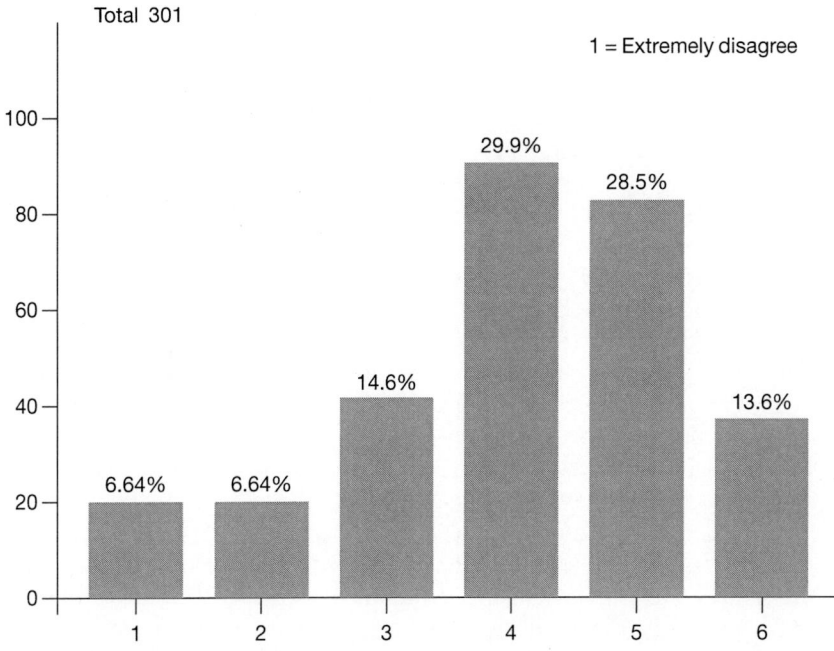

Figure 5.13 If your child had a genetic disease, would you feel responsible for it?

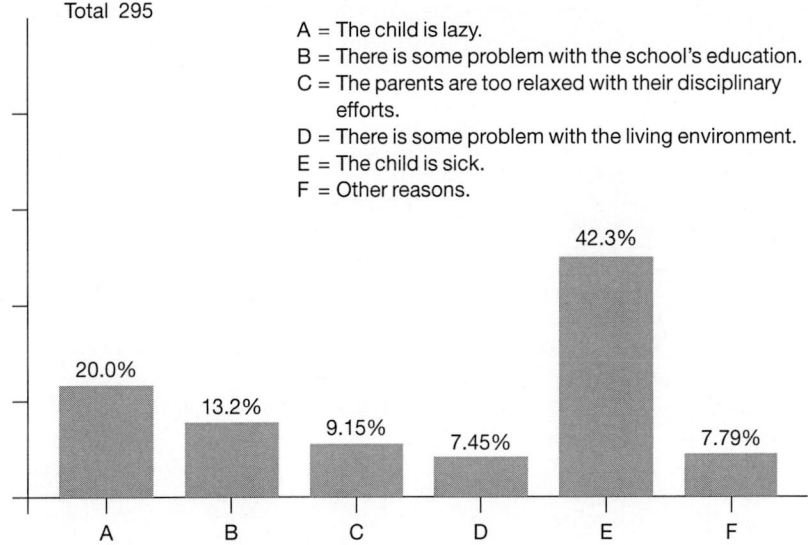

Figure 5.14 If a child has a problem reading and writing, what do you think might be the reason for the problem?

21 Do you think that by understanding the genes responsible for difficulties in reading and writing, one can help in solving the problems?

Most were not optimistic about whether efforts spent in understanding the genes responsible for difficulties in reading and writing would help in solving the problems. Twenty-eight respondents did not feel that by understanding the genes responsible for difficulties in reading and writing, one could help in solving the problems, and only three respondents answered yes for this question. Many did not know the answer and did not respond. Some felt that replacement of the dyslexic gene by a 'better' version of the gene could solve the problem. Four respondents felt that practice makes perfect and dyslexic children should just practise reading and writing more in order to improve their abilities. Some felt that special therapy, education, counselling of the children, and special training of teachers could improve the situation of these children.

22 Do you have any solution for difficulties in reading and writing, and other suggestions for dyslexia and DNA biobanking?

Most did not write any comments for this open question. This might be due to the fact that a lot of people did not feel inclined to type in their responses when they were answering the questionnaire on the computer. Some were suspicious of the benefit of this activity and even said that 'biobanking may result in serious criminal activities, especially in this age of terror attacks'.

Note

The correlation between age and education level was done by Pearson correlation test (2-tailed), and significance level was at the 0.01 level (**) or 0.05 level (*).

References

Antequera F. and Bird A. (1993) Number of Cpg islands and genes in human and mouse, *Proceedings of the National Academy of Sciences of the United States of America*, 90: 11995–9.

Biesecker B.B., Boehnke M., Calzone K., *et al.* (1993) Genetic-counseling for families with inherited susceptibility to breast and ovarian-cancer, *Journal of the American Medical Association*, 269: 1970–4.

Binder J.R., Frost J.A., Hammeke T.A., *et al.* (1996) Function of the left planum temporale in auditory and linguistic processing, *Brain*, 119: 1239–47.

Boggian A.K. and Barret M. (1985) Performance and motivational deficits of helplessness: the role of motivational orientations, *J Pers Soc Psychol*, 49(6): 1753–61.

Chan D., Ho C.S.H., Tsang S.M., *et al.* (2004) Screening for Chinese children with dyslexia in Hong Kong: the use of the teachers' behaviour checklist, *Educational Psychology*, 24: 811–24.

Chan D.W., Ho C.S.-H., Tsang S.-M., *et al.* (in press) Prevalence, gender ratio, and gender differences in reading-related cognitive abilities among Chinese children with dyslexia in Hong Kong, *Education Studies*.

Chan D.W., Ho C.S.-H., Tsang S.-M., *et al.* (under review) Incidence of developmental dyslexia in Hong Kong: What differences do different criteria make?

Chen L. and Woo S.L.C. (2005) Complete and persistent phenotypic correction of phenylketonuria in mice by site-specific genome integration of murine phenylala-nine hydroxylase cDNA, *Proceedings of the National Academy of Sciences of the United States of America*, 102: 15581–6.

Cristiano R.J., Smith L.C. and Woo S.L.C. (1993) Hepatic gene-therapy – adenovirus enhancement of receptor-mediated gene delivery and expression in primary hepato-cytes, *Proceedings of the National Academy of Sciences of the United States of America*, 90: 2122–6.

Goldberg Y.P., Nicholson D.W., Rasper D.M., *et al.* (1996) Cleavage of Huntingtin by apopain, a proapoptotic cysteine protease, is modulated by the polyglutamine tract, *Nature Genetics*, 13: 442–9.

Grigorenko E.L. (2003) The first candidate gene for dyslexia: turning the page of a new chapter of research, *Proceedings of the National Academy of Sciences of the United States of America*, 100: 11190–2.

Grigorenko E.L. (2005) A conservative meta-analysis of linkage and linkage-associa-tion studies of developmental dyslexia, *Scientific Studies of Reading*, 9: 285–316.

Grigorenko E.L., Wood F.B., Golovyan L., *et al.* (2003) Continuing the search for dyslexia genes on 6p, *American Journal of Medical Genetics Part B – Neuro-psychiatric Genetics*, 118B: 89–98.

Ho C.S.H. and Fong K.M. (2005) Do Chinese dyslexic children have difficulties learning English as a second language? *Journal of Psycholinguistic Research*, 34: 603–18.

Ho C.S.H., Chan D.W.O., Lee S.H., *et al.* (2004) Cognitive profiling and preliminary subtyping in Chinese developmental dyslexia, *Cognition*, 9: 43–75.

Hoeyer K. (2005) Studying ethics as policy – the naming and framing of moral prob-lems in genetic research, *Current Anthropology*, 46: S71–S90.

Kan Y.W., Lee K.Y., Furbetta M., *et al.* (1980) Polymorphism of DNA -sequence in the beta-globin gene region – application to prenatal-diagnosis of beta-0-thalassemia in Sardinia, *New England Journal of Medicine*, 302: 185–8.

Karason A., Gudjonsson J.E., Jonsson H.H., *et al.* (2005) Genetics of psoriasis in Iceland: evidence for linkage of subphenotypes to distinct loci, *Journal of Investigative Dermatology*, 124: 1177–85.

Kazazian H.H., Dowling C.E., Hurwitz R.L., *et al.* (1992) Dominant thalassemia-like phenotypes associated with mutations in exon 3 of the beta-globin gene, *Blood*, 79: 3014–18.

Li J.K., Ng M.C., So W.Y., *et al.* (2006) Phenotypic and genetic clustering of diabetes and metabolic syndrome in Chinese families with type 2 diabetes mellitus, *Diabetes and Metabolism Research Review*, 22: 46–52.

Mak J.C., Leung H.C., Ho S.P., *et al.* (2006) Analysis of TGF-beta(1) gene polymor-phisms in Hong Kong Chinese patients with asthma. Homozygous L-SIGN (CLEC4M) plays a protective role in SARS coronavirus infection. Stiff child syndrome with mutation of DYT1 gene. Phenotypic and genetic clustering of diabetes and metabolic syndrome in Chinese families with type 2 diabetes mellitus. A new haplotype of PDCD1 is associated with rheumatoid arthritis in Hong Kong Chinese, *Journal of Allergy and Clinical Immunology*, 117: 92–6.

Molfese D.L. (2000) Predicting dyslexia at 8 years of age using neonatal brain responses, *Brain and Language*, 72: 238–45.

Nicolson R.I. and Fawcett A.J. (2005) Developmental dyslexia, learning and the cerebellum, *Journal of Neural Transmission – Supplement*, 19–36.

Rumsey J.M., Nace K., Donohue B., *et al.* (1997) A positron emission tomographic study of impaired word recognition and phonological processing in dyslexic men, *Archives of Neurology*, 54: 562–73.

Samuelsson S. and Herkner B. (2003) Reading and writing difficulties among prison inmates: a matter of experiential factors rather than dyslexic problems, *Scientific Studies of Reading*, 7: 53–73.

Samuelsson S., Gustavsson A., Herkner B., *et al.* (2000) Is the frequency of dyslexic problems among prison inmates higher than in a normal population? *Reading and Writing*, 13: 297–312.

Senn O., Russi E.W., Imboden M., *et al.* (2005) Alpha(1)-antitrypsin deficiency and lung disease: risk modification by occupational and environmental inhalants, *European Respiratory Journal*, 26: 909–17.

Shaywitz B.A., Shaywitz S.E., Pugh K.R., *et al.* (2002) Disruption of posterior brain systems for reading in children with developmental dyslexia, *Biological Psychiatry*, 52: 101–10.

Simos P.G., Fletcher J.M., Bergman E., *et al.* (2002) Dyslexia-specific brain activation profile becomes normal following successful remedial training, *Neurology*, 58: 1203–13.

Svensson I., Jacobson C. and Lundberg I. (2000) The nature of reading and writing difficulties among inmates in institutions for compulsory care, *International Journal of Psychology*, 35: 164–5.

Syrjakoski K., Kuukasjarvi T., Waltering K., *et al.* (2004) BRCA2 mutations in 154 Finnish male breast cancer patients, *Neoplasia*, 6: 541–5.

Taylor J.M., Dozy A., Kan Y.W., *et al.* (1974) Gene deletion as cause of alpha thalassemia – genetic lesion in homozygous alpha thalassemia (hydrops fetalis), *Nature*, 251: 392–3.

Vulliamy T., Mason P. and Luzzatto L. (1992) The molecular-basis of glucose-6-phosphate-dehydrogenase deficiency, *Trends in Genetics*, 8: 138–43.

Vulliamy T., Beutler E. and Luzzatto L. (1993) Variants of glucose-6-phosphate-dehydrogenase are due to missense mutations spread throughout the coding region of the gene, *Human Mutation*, 2: 159–67.

Wertz D.C., Fanos J.H. and Reilly P.R. (1994) Genetic testing for children and adolescents – who decides, *Journal of the American Medical Association*, 272: 875–81.

Woo S.L.C., Chandra T., Stackhouse R., *et al.* (1982) Analysis of phenylketonuria (PKU) by cloning and characterization of the human phenylalanine-hydroxylase gene, *American Journal of Human Genetics*, 34: A168.

Woo S.L.C., Lidsky A., Chandra T., *et al.* (1983) Prenatal-diagnosis and carrier detection of classical phenylketonuria by cloning and characterization of the human phenylalanine-hydroxylase gene, *Clinical Research*, 31: A479.

Bibliography

Altshuler D., Brooks L.D., Chakravarti A., *et al.* (2005) A haplotype map of the human genome, *Nature*, 437: 1299–320.

Anderlik M. (2003) Commercial biobanks and genetic research: ethical and legal issues, *American Journal Pharmacogenomics*, 3: 203–15.

Anderlik M.R. and Rothstein M.A. (2001) Privacy and confidentiality of genetic information: What rules for the new science? *Annual Review of Genomics and Human
Genetics*, 2: 401–33.

Andrawiss M. (2005) First phase of HapMap project already helping drug discovery,
Nature Reviews Drug Discovery, 4: 947.

Austin M.A., Harding S.E. and McElroy C.E. (2003) Monitoring ethical, legal, and
social issues in developing population genetic databases, *Genetics in Medicine*, 5:
451–7.

Collins F.S., Guyer M.S. and Chakravarti A. (1997) Variations on a theme: cataloging
human DNA sequence variation, *Science*, 278:1580–1.

De Paoli P. (2005) Biobanking in microbiology: from sample collection to epidemiology, diagnosis and research, *Fems Microbiology Reviews*, 29: 897–910.

Duyao M., Ambrose C., Myers R., *et al.* (1993) Trinucleotide repeat length instability
and age-of-onset in Huntington's disease, *Nature Genetics*, 4: 387–92.

Foster M.W. (2005) Ethics watch – genetics and social identity after the HapMap,
Nature Reviews Genetics, 6: 879.

Freimer N. and Sabatti C. (2004) The use of pedigree, sib-pair and association studies
of common diseases for genetic mapping and epidemiology, *Nature Genetics*, 36:
1045–51.

Gibbs R.A., Belmont J.W., Hardenbol P., *et al.* (2003) The International HapMap
Project, *Nature*, 426: 789–96.

Glazier A.M., Nadeau J.H. and Aitman T.J. (2002) Finding genes that underlie complex
traits, *Science*, 298: 2345–9.

Godard B., Marshall J., Laberge C., *et al.* (2004) Strategies for consulting with the
community: the cases of four large-scale genetic databases, *Science and Engineering
Ethics*, 10: 457–77.

Hodgson S.V. and Popat S. (2003) Polymorphic sequence variants in medicine: a challenge and an opportunity, *Clinical Medicine*, 3: 260–4.

Kerem B.S., Rommens J.M., Buchanan J.A., *et al.* (1989) Identification of the cystic-
fibrosis gene – genetic-analysis, *Science*, 245: 1073–80.

Maojo V. and Martin-Sanchez F. (2004) Bioinformatics: towards new directions for
public health, *Methods of Information in Medicine*, 43: 208–14.

Pearson H. (2004) Summit calls for clear view of deposits in all biobanks, *Nature*, 432:
426.

Rommens J.M., Iannuzzi M.C., Kerem B.S., *et al.* (1989) Identification of the cystic-
fibrosis gene – chromosome walking and jumping, *Science*, 245: 1059–65.

Snell R.G., Macmillan J.C., Cheadle J.P., *et al.* (1993) Relationship between trinucleotide repeat expansion and phenotypic variation in Huntington's disease, *Nature
Genetics*, 4: 393–7.

Thorisson G.A., Smith A.V., Krishnan L., *et al.* (2005) The International HapMap
Project web site, *Genome Research*, 15: 1592–3.

Wichmann H.E. (2005) Genetic epidemiology in Germany – from biobanking to
genetic statistics, *Methods of Information in Medicine*, 44: 584–9.

Part II

Biobanking and problems of inequality

6 Human genetic sampling in Indonesia

The interplay between biosocieties and non-biosocieties

Margaret Sleeboom-Faulkner

Africa may be the cradle of mankind, but Southeast Asia has by far the most isolated ethnic groups spread over a huge number of islands in the Indian and Pacific Oceans and is as such the most important reservoir of human genetics, virtually untapped to date.

(Dr X, August 2002, http://www.ihdreg.com/mainpage.htm)

Sanbe Farma has already started setting up the satellite laboratories, and all of them will be running toward the end of the year, equipped with a biologist and a physician plus following, and an adequate budget to start off with: mainly aimed at collecting DNA and families.

(Dr X, 23 February 2002, transl. MS-F)

Introduction

On his quest to place Indonesia on the genetic world map, a Dutch scientist, Dr X,[1] spread the word that he would set up the largest biomedical research centre in Indonesia, using blood samples gathered from various laboratories, medical centres and the jungle. He would study genes for familial syndromes, such as cancers and forms of mental retardation, some of which are considered to be specific for certain ethnic groups in Indonesia.

PT Sanbe Farma, a manufacturer of pharmaceuticals in Bandung (West Java) in 2001 announced its decision to start a registry of inherited diseases of the Indonesian people in all her genetic diversity. The project was advertised as 'a stimulus for future Central Government initiatives for a Human Genomics program to be established in this vast country of more than 220 million people and its enormous genetic resources' (cf. www.sanbe-farma.com/biotech.html, 8 February 2006). Its leader was Dr X.

Much has been written on the targets of the Human Genome Diversity Project (HGDP) (Cavalli-Sforza and Cavalli-Sforza 1995; Reardon 2001:365), but the incentives for and motives of geneticists to set up human genetic biobanks as expatriates in 'exotic' societies are little understood. What happens when a 'Western' scientist installs himself as a genetic sample collector in a country that has not developed the institutions associated with 'biosocieties' such as 'biological citizenship' (Heath *et al.* 2003; Rose and Novas 2003), a system of intellectual property rights and the legal and political instruments to regulate such activities, seems to be open to many interpretations. Various attempts have been made to theoretically 'frame' the practices of human geneticists as imperialist (Desowitz 2002; Harry *et al.* 2000; Marshall 1997:565; Rifkin 1998; cf. Lock 1994:603–6; cf. Macer 1997), racist (Gordimer 2002; Lewontin 2001:1–40; Reardon 2004) and positivist (Carlson and Stimeling 2002), or to hail them as great explorations (Avise 2001; Stock 2003; Sykes 2002) and progress for mankind (Harris 1998; Savalescu 2001). Others have related the creation of genetic databases to the creation of 'imagined communities' (Busby and Martin: 2006). The problem is, however, that such general approaches hardly take into account the nature of the 'receiving society', which may not have developed the institutional infrastructure of a 'biosociety'. If its people have not developed any form of biosocial identity (Rabinow 1992), have no means to reflect upon the financial and political implications of cooperating in the creation of a registry of biologically inheritable diseases, and do not translate the collection of genetic information into a resource of profit, science, and public health, it is clear that this 'non-biosociety' is at a disadvantage when targeted for genetic sampling practices: the general public does not have the financial means and conceptual tools to negotiate conditions for fair cooperation.

In this chapter, the author takes the view that whether we judge Dr X to be a pioneer of science, an imperialist exploiter, or a well-meaning dreamer, it is no more relevant to genetic sampling activities than to any other commercial undertaking. Instead, the materials will be used to make the following arguments. First, genetic technologies developed in so-called biosocieties, characterised by identities, rights and skills associated with biological citizenship (Rose and Novas 2003) have a different impact on 'non-biosocieties' than they have in biosocieties. Second, taking advantage of the different biosocial conditions is unfair in the Rawlsian sense, as no principles of justice are agreed to in an initial situation that is fair (Rawls 1973:12–13). Third, the host country becomes a 'pre-biosociety', a society that is more or less forced to adopt protective regulatory measures, following the regulatory concepts developed by biosocieties. And, finally, the author tries to make plausible the idea that any creation of unenforceable regulations and guidelines accompanied by new categories of genetic identity in such a 'pre-biosociety' could aggravate already serious ethnic clashes and nationalistic sentiments.

A main aim of this chapter is to understand the relationship between the possible motivations of a scientist to bank human genetic data and the institutional context in which the banking takes place. The author does this on the basis of e-mail correspondence that took place between Dr X and a group of people that he involved in his plans to set up an HGDP in Indonesia over the years 2002–2003. Much of this correspondence was copied to a group of people, and one researcher invited to participate in Dr X's project, pointed out to this author its relevance to discussions on 'Western' views of science and its practices in 'non-Western' cultures. By analysing Dr X's e-letters to fellow scientists, academics and politicians, some of whom this author shall refer to by using their initials only,[2] the author will provide six perspectives on Dr X's plans to motivate his work on genetics to Indonesia.

The question is what these perspectives contribute to a better understanding of genetic sampling practices. This question the author tries to answer in the second part of the article (An institutional approach, p. 130), where the institutional embedding of the Dr X case is discussed. It is claimed that the validity of the theoretical perspectives used to analyse the Dr X case depends on whether they sufficiently take into account the institutional embedding. Such an approach to genetic sampling should incorporate an assessment of socio-economic inequalities, both on an international and national level, public healthcare needs, research regulation, and differences in academic cultures from a comparative perspective. The second half of this article then explores the implications of the institutional environment for the introduction of new genetic technologies in Indonesia and it will make some observations on the use of the paradigms discussed.

Research problems

Dr X, having been offered a job as director of the Biotechnology section in Sanbe Farma (est. 1975), the largest pharmaceutical company in Indonesia,[3] decided to take a chance and move to Bandung in West Java. Finishing a respectable career as a scientist and academic at the Free University of Amsterdam at 60 years of age, and having built up a publication record exceeding 300 articles published in peer-reviewed scientific journals, Dr X decided to start a new career. It would be a true new beginning. Not so much because it required his relocation in another part of the world, but rather because it meant a new challenge among people who would appreciate him and who would not put a myriad of bureaucratic guidelines and rules in the way of getting on with science. And last but not least, his new post would allow him to realise his dream of placing Indonesia on the world's human DNA map. A homepage called 'The Indonesian Hereditary Disease Registry' would offer to the whole world of human geneticists:

> the analysis of the germline mutations involved in all kinds of diseased families, in a country with 200 different extended families or 'ethnic groups' if you wish to call them in that way. A dream to come true, or a nightmare? Only God knows, but I am working day and night to realise it and things are moving at great and unexpected speed already. [15 February 2002, http://www.ihdreg.com/mainpage.htm]

What is so problematic about setting up a hereditary disease register? Dr X's concerns were not of the bioethical kind – he has an optimistic, constructive approach. For Dr X, as head of the Division of Biotech and Research of Sanbe Farma, the main problems are in the following areas:

- *Human resources:* Here the concept of human resources refers to sampling targets. Prospects for genetic sampling in Indonesia look better than in the Netherlands.
- *Research funding:* Funding is needed for the purchase of expensive reagents. Funding from the Netherlands and relatively cheap scientific labour in Indonesia could utilise the economic and anthropological differences in the Netherlands and Indonesia.
- *Data collection:* Blood samples are needed from ethnic groups, including each family's mDNA profile and a Y-chromosome set of DNA markers to verify their ethnicity.
- *Institutional limitations:* The academic bureaucracy and research regulation in the Netherlands set institutional limitations to scientific possibilities. At Sanbe Farma Dr X enjoys relative freedom in setting up his research and receives respect from his colleagues and students.

For Dr X to solve his problems, a scientific attitude, an entrepreneurial spirit and the ambition to generate knowledge in the biomedical field seem to be essential on his path to scientific and human progress.

Progress

This section tells the story of Dr X's plans and activities by regarding him as progressive.

Scientist?

Dr X has various scientific aspirations and worries in the fields of genomics and biomedicine in general, and in the fields of DNA collection and molecular biology specifically.

Interested in the field of genomics, Dr X on 19 February 2002 proposed to people associated with the Dutch Genomics Initiative (NGI), such as its director Peter Folstar, Emil Broesterhuizen[4] and then-Minister of Education Louk Hermans, to set up an Indonesian Satellite of the Dutch 'Genomics Institute', together with the Inter University Centre (IUC) Biotech of the Institut Teknologi Bandung (ITB, est. 1959) in Bandung. In Indonesia the ITB is highly rated in the field of medically oriented molecular biology. Dr X's own research group at Sanbe Farma had started an ambitious programme to type potentially new human papillomaviruses in Indonesia. Dr X regarded it as a prelude to an adequate vaccine against the development of cervical cancer, the major cancer problem in women in Indonesia.

Dr X's scientific endeavours were clearly linked to the aims of the so-called Human Genome Diversity Project (HGDP, discussed below, p. 125), as his challenge was 'to put Indonesian ethnicity onto the world DNA map' in general, and to apply 'DNA mutational analyses of germline mutations of the relevant genes for familial syndromes present in Indonesia' in specific. Some of these mutations, according to Dr X, may actually be particular to ethnic groups in Indonesia and in relation to ethnicity to be determined from mitochondrial DNA mutations[5] [19 February 2002]. Thus, Dr X proposed a programme for collecting DNA samples with genetic abnormalities of all of the, over 200 [*sic!*],[6] ethnic groups in Indonesia. Needing a map that gave him an indication of where to start collecting DNA samples, Dr X directly began to look for anthropologists. His main biomedical concerns here were the genetic deviations for familial cancer and mental retardation. All pedigrees were to be provided on the Web with the request put to the international community of geneticists to assist with the detection of germline mutations.[7]

Close to his heart was the formulation of new policies in the medical/pharmaceutical field. On the basis of his experience of several years as a lecturer at Gadjah Mada, what he called a 'very poor setting for science', followed by over a year in one of the richest settings, Dr X listed the basic principles he believes should govern a new law, called the Medical-Pharmaceutical [*sic*] Research Law (MPRL):[8]

1 Medical doctors are surrogate 'tax' collectors from the poor in their private practices. If they are employed in universities/institutes of learning they tend to disregard science and progress in science as irrelevant and neglect their duties as 'scientists', so they must pay 'research tax' to make up for lack of duty (unless they can 'prove' by scientific papers and science citation index that they are fulfilling their scientific duties).

2 Pharmaceutical companies are surrogate 'tax' collectors from the poor, too, and they should invest in new methodologies and advance medical and pharmaceutical technologies by investing in it. If they don't do it they have to pay research tax; if they do, they can get tax breaks on investments.

3 A new medical/pharmaceutical research committee should get the instructions, means, etc. to measure scientific output based on SCI and other parameters of the medical doctors in universities and the pharmaceutical companies. A new law should be prepared with the ultimate goal of propelling Indonesia into the modern age of medical biotechnology.

4 Goal: Get Indonesia from place 79 to place 40 in 5 years' time in terms of scientific output.

Dr X considered himself to be a scientist at heart, which he expresses on a webpage of Sanbe Farma as follows:

I close with a poem hoping to convey to you how I feel at the beginning of a new career, now as a "businessman", but still of course a scientist at heart:

> Science is like a cast,
> Deep, and living ocean
>
> Science is a world
> Full of thought and emotion
>
> Being a scientist is part of
> Mankind's greatest past,
> I reveal for the future a vision to last.

Businessman?

Dr X expressed his entrepreneurial interests in his fund raising, planning and brokerage skills. For instance, he wanted to help the IUC, an internationally competitive institute in the field of molecular biology. Though it had highly qualified researchers and equipment, it could not afford the expensive reagents.[9] In order to solve this problem, Dr X explained:

> I am actually planning to build the largest Indonesian Biomedical Research Center in Padalarang in a few years and I am sure that ITB Rektor 'Kus' [Rector Kusmayanto of ITB] and my Pak Jahja [Dr X's 'boss' at Sanbe Farma] are willing to share the burdens of such new facilities to be built. [19 February 2002]

Apart from the central laboratory in Bandung, Dr X also has a laboratory in Jakarta (Dharmais) and Medan (USU). Additionally, he was building new ones in Semarang, Yogyakarta, Makassar, Surababaya and Bali/Lombok. The laboratories make DNA from collected blood, needed for scientific projects. It is possible, Dr X argued [21 February 2002], to collect the blood on blotting paper in the jungle, dry it in the air, wrap in a plastic envelope and send it off to a laboratory where the DNA could be purified. Dr X then recommended colleagues from the Netherlands [HJ] to come to Indonesia and 'claim' their department at the hospital and to start getting equipment together. Other experts he invited to move into his microbiological and diagnostic laboratories.

Always on the look-out for lucrative opportunities, Dr X was keen to approach key persons for funding. Thus, when Dr Henk Smits of the Royal Institute for Tropical Research (KIT, Amsterdam)[10] during his visit in March offered assistance, Dr X was prepared to 'take on anything that can generate medical progress in this country'. And on the occasion of Smits' official speech, he announced, 'we will have collected all established microbiologists and we will have invited a lot of them from other parts of Indonesia' [21 February 2002]. Similarly, while addressing Kusmayanto, the Rector of ITB (whom he addresses familiarly as 'Kus'), he announced:

I am bombarding the Minister of Education Hermans with messages hoping that he will play Sinterklaas [Father Christmas] towards us, as all good Ministers can do at the end of their term if they have saved some money. [21 February 02; transl. MS-F]

Similarly, to Peter Folstar, director of the NGI ('Dear Professor Folstar'), Dr X explained that the difficult job of analysing germline mutations could take place in Indonesia by using cheap labour and careful planning. The main work could be done at ITB, while Sanbe Farma could make use of one of its MOUs to test all DNA containing adequate information on 'first right of refusal'.[11] To make Sanbe Farma more attractive, Dr X offered to look for other jobs related to molecular biology, varying from the genomes of mammals to those of bacteria. He also asked for advice about whom to contact in universities and Departments of Human Genetics in the Netherlands, suggesting half-jokingly 'a little syndrome for each?' An anthropologist was meant to make an initial map of the tribes of Indonesia to facilitate the search for genetic deviations in the 'jungle'. Dr X was full of confidence:

My laboratory and the number of personnel are growing so fast that I can hardly keep up with it myself. God, now I realise how hard it has been to keep body and soul together in the Netherlands (but nevertheless [I had] 330 publications Gentlemen, so it could not have been me [!])

Dr X expressed his faith in PT Sanbe Farma, compared to other pharmaceutical companies: 'Over the last decades, the over 200 pharmaceutical companies (that have hardly any research and patents to speak of) in Indonesia have meant nothing to science.' The newly available funding at PT Sanbe Farma for genetic research therefore, he boasted, causes amazement all around, especially considering the high amount, which he estimated to be over a million dollars and growing. [24 June 2002, to CB]

Enlightened?

Dr X has enlightened ideas about improving the world, especially where poor countries on the road to progress are concerned. He acknowledged that his plans for the registry 'may not be as pressing to the Indonesian government as are problems of infectious diseases, such as tuberculosis, malaria, dengue and emerging venereal diseases (ranging from HIV-induced syndromes to HPV-induced cervical cancer)'.[12] Dr X realised there was a need to attract research funding from elsewhere. Although the thalassaemias have received attention from the Eijkman Institute for Molecular Biology,[13] and Fragile X syndrome and BRCA1/2 have also attracted research funding, other genetic disorders were understudied and diagnostic resources were sorely lacking. To remedy this situation, Dr X tried to attract attention from Broesterhuizen to the importance of the education of researchers in Indonesia:

Continuous transfer of know-how through the young professionals is of course a must: 'A Progress in Methods is a Progress in Science is a Progress in Personal Development' should be the motto. [19 February 2002]

And when Dr X did not receive the desired support, realising that his plans did not fit the purpose of the Dutch Genomics Initiative funding, he was pleased that director Folstar at least expressed his willingness to look for indirect channels through aid for education.

Dr X also stimulated researchers such as HJ and BT – 'who both were born in Surabaya' – to visit his institute:

to brainstorm about how to help Indonesia, beginning the reparation of ties between two peoples that after all have been connected for centuries, a feeling that is still alive and kicking.' [21 February 2002; transl. M-SF]

These forms of aid and 'people-bonding' were not confined to the historical relation and blood bonds between Indonesia and the Netherlands, but extended to possibilities for the people of Southeast Asia. With the promulgation of the Asian Free Trade Area (AFTA) in 2003, Dr X announced that:

specialists from elsewhere can also come here to exercise their trade: borders will be opened and foreign competition for the Indonesian Medical People is on its way. About time, too. [21 February 2002; transl. MS-F]

Dr X expressed his belief that 'Perhaps this Program can be considered an "answer" from the poor world towards the Human Genomics Program – carried out by the affluent world'.[14] PT Sanbe Farma was willing to meet at least some of these costs for ethnic groups in the various parts of the country if it could take responsibility for the storage of DNA and the pedigree records to be displayed on the Web. 'This', Dr X emphasised,

only in order to convince the Government eventually to take responsibility and create an Indonesian Human Genomics Program based on a modern resource of genetic material of humans and their diseases, a modern 'garden' like Bogor, with its huge resources of genetic material of plants from this island-rich vast Archipelago and its jungles, available for every serious scholar in this world. [1 August 2002]

In short, providing scientific aid, building bridges, creating possibilities and helping poorer nations in general were activities that Dr X, in his quest for support, held in high regard.

Exploitation?

This section tells the story of Dr X's plans and activities by regarding them as exploitative. It makes use of the insights of perspectives that critically observe genetic reductionism where others see scientific progress; they see imperialism in action where others see creative entrepreneurship and global free trade; and they spot Orientalist relations of inequality where others see enlightened dedication and mutual help. These critical perspectives are used *heuristically* as a basis for the main argument in the section below (An institutional approach, p. 130) in favour of an institutional approach that tries to understand the transnational interstices or gaps between regulatory, educational and welfare provisions in modern biosocieties and in a pre-biosociety on the road to developing notions of biological citizenship.

Genetic reductionism?

As part of the Human Genome Diversity Project (HGDP), samples from indigenous populations, according to Luca Cavalli-Sforza, its initiator, human genetic data collections serve to avoid the irreversible loss of precious genetic information (Cavalli-Sforza *et al.* 1991:490–1). The identification of genetic difference between ethnic groups (such as a different gene sequence that causes immunity to a disease) is also believed to have human (medical, cultural and historical) and economic value. This perceived value has resulted in a new 'gold rush' where universities, governments, corporations and private researchers are seeking to identify human genetic variation among remote populations (Marshall 1997:565). Dr X clearly also recognises their value:

> Southeast Asia with its more than 1000 ethnic groups, especially in the island-rich nations of Indonesia and the Philippines, many isolated since time immemorial and therefore unique founder gene mutations is in this respect one of the best genetic resources on Earth, probably superior to Africa, where mankind evolved and where genetic differences are the results of the longest evolutionary time, but where isolation of extended 'families' or 'ethnic groups' is clearly at a smaller scale. [August 2002]

The initial conceptualisation of the HGDP has been widely criticised for regarding indigenous peoples as mere research subjects, with little concern for their continued livelihood. Various ethnic groups have protested against population genetics, which they claim has been conducted without prior consultation of either the individual and the entire community involved (cf. see Knoppers 2003; Peterson 2001; Pilnick 2002; Sándor 2003; Sutrop 2004). Because the HGDP was planning to collect blood samples, some groups called the HGDP the 'Vampire project' (Lock 1994:603–6), while others were angry because they believed that they were possible target populations even though no community representatives had been contacted about the project.[15]

While criticism of the HGDP was growing, plans for a so-called international haplotype map (the HapMap) were gaining support. It was hoped that commercial and public efforts to assemble DNA polymorphisms would contribute to the discovery of SNPs that affect gene function.[16] Most successes in finding genes that contribute to disease risk have been for rare diseases caused by single genes, such as cystic fibrosis.[17] And it is these single gene disorders that are the ones that Dr X proposes to register. But as shown by genetic research on individuals of Ashkenazi Jewish descent and African-Americans, concerns are raised about the augmentation of existing prejudices and genetic pretexts for discrimination in situations of conflicting political and socio-economic interests.[18] Considering its history of ethnic strife and discrimination of the mentally handicapped, Indonesia may be no exception.

These discussions seem to have escaped Dr X entirely. Apart from asking an anthropologist to lecture on 'The Ethnic Groups of Indonesia: traditional and molecular anthropology' that could serve as a blueprint or ethnic map, Dr X tried to obtain information from Professor Sangkot Marzuki, a molecular geneticists involved in mDNA research:

> Perhaps it is published already and it is likely that I now need to have the mDNA profile of every family and possibly even a Y-chromosome set of DNA markers to verify ethnicity. But before I have set it up, naturally, I can deduce from the name etc. etc. which group I'm dealing with.[19] [Transl. MS-F]

Of course, other factors also contribute to population differences, such as culture, language, environment and lived experience. If these factors are neglected in the definition of populations, one may ask if haplotyping would privilege some group definitions over others. What would be the effect of the creation of new socio-genetic boundaries between populations?

Imperialism?

Bioethical and financial concerns exist about access to genetic data in registers, which socio-economic groups will be able to turn these data into profit, and which genetic diseases will be the focus of research. It was mainly such concerns that led Asian nations in November 2004 to set up their own haplotyping initiative.[20] Dr X was also concerned. Although he would share genetic data on the internet, on the registry website Dr X announced:

> Our Registry is in the beginning [phase] and we plan to present all relevant family information on the Web. Requests for help in gene mutation searches will be given only if the pedigree is as complete as possible, if relevant medical records are sufficient, if enough DNA is present to send

out aliquots to interested groups worldwide and if we are not doing the analysis ourselves [1 August 2002]

In other words, the registry is offering help in exchange for research and medical data. It becomes clear that providing information on founder genes on the Web is one way of getting others to help.

> Once published, specific data on the kind of mutations, whether it is a Founder gene mutation of a certain ethnic group etc. will also be revealed on our Web site and submitted to McKusick's 'Online Mendelian Inheritance in Man' (OMIM) database, *so as to inform all interested human geneticists in the best and fastest possible way in order to generate their maximum interest to help us using their resources.* (1 August 2002, http://www.ihdreg.com/mainpage.htm, [*italics* are mine MS-F])

Proposing to Dutch Genomics director Peter Folstar that Sanbe Farma sample and test the DNA of Indonesian families, Dr X offered Dutch scientists access to obtain the scarce information on the genome of families with a genetically unique make-up.

> Human genetics in the Netherlands is gradually developing a shortage of families and therefore would find here a fantastic hinterland. A lot of interest exists at for example the VU [Free University, Amsterdam] and the KUN [Royal University of Nijmegen], who already have started joint research projects. [Transl. MS-F]

Here Dr X seems to be simply playing on Dutch anxieties about missing out on good opportunities to access valuable human research materials, and consequently research publications, fame and funding.

Despite his plans for building laboratories and gaining access to the DNA samples of families from hundreds of ethnic groups, a problem of sample shortages developed:

> For as yet I do not even have 10 families here in Bandung (I heard from Gerard that you have 23 from Yogyakarta) and I am a little ashamed about my big mouth, but I can assure you that the entire process of looking for families has begun and is progressing faster and faster. Before the end of the year I will certainly have over a 100 families and the number will accelerate exponentially when the satellite laboratories are running properly (until now only a few families from Medan), but in Semarang (Dr Sultana Hussein, a genetics star in the Indonesian firmament, with which we cooperate closely and have given much money to) there is an entire school (Semin) with Fragile-X syndrome,[21] as you know, Ben. [23 February 2002; transl. MS-F]

The bottom line seems to be money after all:

> Ah, before I forget: we can employ 20 biologists here for the same price of
> one over there with you. And I can assure you that they are at least as
> smart. And once they get the chance to work they gain experience fast
> enough, and we work here until Saturday afternoon. [23 February 2002;
> transl. MS-F]

Orientalist?

According to Edward Said, Orientalism is a 'Western style for dominating,
restructuring, and having authority over the Orient'. That authority was the
ability to say what was and was not 'Oriental'. This usually includes:

1   An inversion of what is thought to be European civilisation, justifying
    attempts to enlighten the backward Other;
2   taking over values of the Oriental elite, gaining a foothold in the Orient by
    supporting the local elites;
3   an inversion of history that recognises the Orient as a richness of historical
    resources). (Cf. Said 1979:1–9)

As a generic term, Orientalism refers to distortions in the perception and
analysis of 'Other' cultures. One distortion exaggerates the differences between
the familiar and the strange: a focus on the exotic. Another distortion is treating
a society as though it is an unchanging expression of some basic essence or
genius (essentialism). Here, social and cultural practices and institutions are
portrayed as being 'what they are because they are what they are' ontologically
(so that no empirical material can alter or dislodge the claims made). A third
distortion portrays and analyses Oriental society as though it is radically separated from the West. This occurs especially in works that ignore points of
painful contact between this Other society and the West, ignoring the adverse
effects of colonial relations and ignoring Western intrusions in that society.

In Dr X's plans to set up a joint registry Indonesians and their DNA were the
object of research, but no Indonesian voices were listened to. Is this a coinci-
dence? Reasons for setting up the registry included stimulating local research,
providing medicine, placing Indonesia genetically on a world map, and this
was all done in cooperation with a large pharmaceutical company, Sanbe
Farma. Not only did the registry support the educated elite of Indonesia, but Dr
X's enlightened charity to the poor in the 'jungle' and his attempt to place their
DNA on the World Human Genome Map could be viewed as paternalistic.
Furthermore, just a few years of experience seems to be sufficient to Dr X to
formulate Indonesian science policies:

> In the mean time I am now able to 'formulate' a new Science Policy for
> this Country in the Medical/Pharmaceutical fields after having worked for

several years as Lecturer at Gadjah Mada in a very poor setting for science, followed by over a year now in one of the richest settings for science in this country.

Prompting colleagues to come to Indonesia seems to play on a sense of paternal benevolence:

Compared to the Netherlands 'nothing is finished' here and over there 'everything is finished'. You are no longer really needed over there, H, but here you are really needed. [to HJ; transl. MS-F]

This benevolence seems to be naturally 'Dutch':

An 'ex-pat' from 'Belanda' is still in peoples' good books, especially with the elderly who immediately start speaking Dutch and are proud they still can. And of course they still need all the help in the world necessary to keep/make this gigantic archipelago inhabitable for hundreds of million people. [21 February 2002, to HJ; transl. MS-F)

Speaking of recreating a bond between the educated elites of the Netherlands and its former colony, Indonesia, Dr X seems blissfully innocent of the atrocities associated with that same history (see Multatuli 1969 [1875]):

I take the liberty of sending this message to the people involved as well as to Minister Hermans whom – or so I wish to think – would not be adverse to such a 'joint venture', re-establishing the links between the respective communities of high learning of two countries which – alas [*sic!*] – had such strong bonds in the past and I am sure strong dreams for a combined future. [19 February 2002, to EB]

His efforts seemed to be those of a misunderstood fighter for a great cause:

Thanks again for your message, which I need once a while to 'air' my thoughts and to 'keep my job'. Even Don Quixote needs recognition, I believe, if only from Sancho Panza, and especially when he is chasing windmills far away from home. Don after all was 'at home' in Spain, was he not. [21 February 2002, to HJ; transl. MS-F]

To his critical colleague, CB, he expressed amazement when CB points out some of the political dangers of his project:

You had a remark on genetic research being initiated in 'all sorts of exotic places' in this world by the pharmaceutical industry, such as Iceland. Do you mean by this that Indonesia is also such an exotic place? And if so in what sense? The population in Iceland is homogenous, there are hundreds here and in the Netherlands probably none or it would be Giethoorn.[22] [24 June 2002, to CB; transl. MS-F)

Did Dr X really not know about the politics and bioethical complications of genetic sampling and setting up databases?

An institutional approach

A closer look at biobanking in the context of Indonesian legal and political institutions makes clear that the six perspectives described above have little to say about the bioethicality of genetic sampling in Indonesia, unless awareness is introduced of the differences between societies that have and those that have not developed notions of bioethicality and regulatory instruments for the maintenance of biomedical data. These differences between societies, in turn, are intimately related with economic development, education, public health policies, and the ability to implement research regulation. In other words, in order to protect itself, a society exposed to advanced biomedical research is more or less forced to develop new notions of bioethicality and biolegality.

The bioethics of health, food and sanitation

Although Indonesia, as one of the eight Asian tigers, in the first part of the 1990s had annual growth of 6 to 8 per cent, the unexpected 'Asian crisis' in 1997–98 turned this into decline. Economic development had a negative effect on the development of Indonesian bioethics in a very basic sense: food prices shot up and the poor – by 1998 about 60 per cent of the population lived below the poverty line – suffered (Bergstrom 2004). Recovery in Indonesia was comparatively slow, as it was complicated by problems accompanying the shift of power at the end of thirty-two years of authoritarian regime under President Sukarno. Although the following 'reformasi' was gradual and patchy, in 2002 the percentage of people living below the poverty line had decreased to 27 per cent.[23] Nevertheless, the GDP in 2002 averaged at just US$994 per capita, the unemployment rate was 9.1 per cent (ADB quoted in Bergstrom 2004:145), and the situation in the areas of sanitation, healthcare and nutrition was poor. Moreover, according to the UNDP Human Development Report 2001, 66 per cent of the population was not using adequate sanitation and 24 per cent of the population was not using improved water resources (1999). The proportion of underweight children under five (1995–2000) was 34 per cent, undernourished children 6 per cent, and under-height children 42 per cent.[24] It should be obvious that Indonesia was not just waiting for high-tech solutions to its healthcare problems.

Nirmal Kumar Ganguly, Director General, Indian Council of Medical Research (ICMR), emphasised the broad healthcare needs:

Against the background of the attainment by all of health rights, achievement of global health equity, increase in healthy life expectancy, and access for all to essential, quality health services the evolution of bioethics needs to be traced to its present concern with the health of populations, and identify some of the most important inequalities that confront contemporary society.[25]

A particular sensitive issue Ganguly defined as follows:

Potential conflict exists between policies aimed at reaching vulnerable groups and those seeking to achieve aggregate health benefits.[24]

The concept of vulnerable groups here refers to those who cannot afford or do not have access to basic healthcare facilities. These are the circumstances one has to think of when 'ethnic families in the jungle' are asked to provide blood samples for a registry on the Web.[26] When institutions essential to daily survival, such as healthcare institutions, social and economic security institutions (through local institutions such as the family or local associations or government institutions) and legal institutions that regulate and overview research and healthcare are not securely embedded in society, vulnerable peoples may be in no position to make a free and informed decision about participation in a project on their 'unique' genetic make-up. The risk of genetic stigma, and the unavailability of financial means to purchase medicine, receive genetic counselling or genetic therapy means that, besides the high risk of physical suffering, they and their families carry a great burden of psychological and material uncertainty. As the development of welfare institutions may take decades, this problem should be viewed as structural.

It was only in the late 1980s that the Indonesian Medical Association formulated ethical guidelines for medical research, but they had little effect. In 1995, government regulation concerning health research and development were put in place. Of special relevance is Chapter IV on health research on human beings, prescribing informed consent, health and safety measures, provision of information on the research in question, and the protection of confidentiality of patient information and records.[27] However, the institutional capacity to enforce compliance to these guidelines is weak. Apart from lacking resources to do so on the part of the government, it has also proven difficult to integrate medical ethics with the extant medical curriculum in universities (Sjamsuhidajat 2004:51–2).

Not until after issuing the 'Guidelines for Good Clinical Practice' by the Indonesian FDA and the translation of the 'WHO Operational Guidelines for Ethics Committees' in 2001 did the Ministry of Health (MOH) in October 2002 establish a National Committee on the Ethics of Health Research (NCEHR).[28] According to a joint study by the MOH and the WHO, twenty-six sub-national ethical review committees were set up in eleven provinces. However, the study also shows that there was no single standard among these various committees with regard to membership, procedure, nor the perception of bioethics in health research. It led to doubts about its effectiveness.[29] Especially, the capacity to oversee clinical research and compliance with guidelines are in need of reinforcement (Slamet and Elengold 2002:77).

The inefficacy of research guidelines is partly related to a poorly functioning bureaucracy. Systemic corruption and a lack of legal certainty can also partly explain the uncompetitive environment for conducting business and the slow economic recovery from the financial crisis in the late 1990s. Economic problems have been added to by recent acts of terrorism, unequal resource distribution among regions, a lack of reliable legal institutions, a weak banking system, and endemic corruption. While traditional sources for financial and legal aid, health and social security have weakened during the period of industrial and urban expansion in the 1980s and 1990s, governmental and non-governmental institutions have not been able to fill this gap (Abdullah and Hüsken 2003). In short, without welfare institutions, effective supervisory organs and a financially secure population, families and individuals do not have the required support to decide about entrusting their DNA samples to biobanking projects.

Institutional mobilisation and Dr X's initiative

In February 2002 Sanbe Farma was flourishing. On a global level, governments were pouring funding into genomics and optimism reigned. Official institutional exchanges between Indonesia and the Netherlands in the areas of biotechnology had started to increase in the 1990s. The visit of representatives of the KNAW and project leaders to Bandung in February 2002,[30] led by Minister Hermans, was one of the results. There were close links with the activities of the Eijkman Institute for Molecular Biology. Sangkot Marzuki, who had led the institute since its re-opening in 1992, was a strong competitor for biotech funding in the areas of molecular diagnostics and genetic sampling.

It was during the 1992 trip of Minister Hermans that contacts were made with Dr X, which aroused his hopes on funding from the planned genomics programme [24 June 2002]. At the time, Dr X was working on a Konigin Wilhelmina Fonds (KWF, Dutch cancer charity) project in Yogjakarta. When Dr X spoke with Minister Hermans, he had already been asked to work for Sanbe. He had just set up a laboratory and employed some academics of ITB, so that it was only natural for Dr X to see fertile grounds for cooperation. In fact, it is the availability of the institutional set up of KNAW and KWF that led to opportunities for them to have the meeting in the first place.

Dr X took every opportunity to mobilise his and other people's institutional networks, ranging from Indonesia and Singapore to the Netherlands. In his e-letter, Dr X relates how he had managed to befriend the Rector of ITB, Kusmayanto, musing that he 'according to tradition will become the next Minister of Research and Technology', and proposed to start preparations for his new 'Medical Pharmaceutical Research Law' (MFRW) law [21 February

2002]. It was of great importance to Dr X not just to be treated with respect but also to be accepted in the company's institutional culture:

> At work they have started to call me Pak Jo now, instead of Dr Jo, after the example of the Rector [Kusmayanto of ITB]. In short, I am accepted as one of them and no longer the odd man out. My young academics and other personnel worship me and it is a pleasure to work day and night. [22 February 2002, to HJ; transl. MS-F]

Dr X feels clearly more 'at home' at Sanbe Farma than at his former laboratory at the Free University (VU, Amsterdam):

> And in fact I can do top research again without any further restrictions, and as well as in my laboratory at the VU: I no longer can see the difference between my laboratory here (over 10 permanent staff work there already) [which] has nearly doubled in size of what I had at the VU and with which I ended [my stay]. Last week, again, I acquired a full-time molecular biologist (yes, promoted and all and top of the nation here) and a full-time pathologist. And one day soon we will purchase a Cytovision caryology system of US$100,000, while we are selecting the Flow Cytometer at the moment. You would not think it possible in Indonesia, would you? (22 February 2002, to HJ; transl. MS-F]

When, in March 2001, the Genome Institute of Singapore (GIS) was set up as the national flagship program for the genomic sciences in Singapore, Dr X had seen an opportunity for sampling activities to traverse Indonesian borders:

> In Singapore a 'Genomic Institute' has just been set up by a Prof Dr Ed Liu,[31] who is pleased to coordinate things with me and with whom I want to observe all of the 500+ Southeast Asian ethnic groups by looking for support in the Philippines, Malaysia, Thailand and the rest of the 'region'. Of course this is something else, something that even entire Europe cannot beat for all of these 'ethnic' groups are just 'large families' of course, who have lived in isolation in this gigantic archipelago for thousands and thousands of years. (23 February 2002, to P en G (VU), B en F (KUN), D (EUR)(via HJ), C (RUG) en G (Ingeny in Goes) [transl. MS-F]

Judging from the 2004 HapMap initiative to haplotype Asian DNA, however, it seems that Marzuki's prestigious institute had already filled in the slot for Indonesia (see Normile 2004:1667).

Chances of Dutch governmental support seemed more promising. Despite his widespread cynicism about the uprightness of Dutch political institutions, Dr X's faith in them is strong enough for him to believe that a reasonable proposal would be judged on its scientific merits. For this reason, Dr X expressed the hope that the Minister for Education, Louk Hermans, would 'play Father Christmas'. After all, Hermans had 'just given 2 million Euros to the charity for sports and children run by "Kus" (Kusmayanto, the Rector of ITB), surely leaving enough to spend some more – e.g., US$80 million before stepping down.' [22 February 2002, to HJ] Through the Minister and the vice-director of the Board of Education and Science, Emil Broesterhuizen, he was introduced to NGI director Peter Folstar, who showed interest in the idea of a joint Indonesian–Dutch genomics institute [23 February 2002]. Although happy to have direct access to the 'highest Dutch genome', it did not diminish Dr X's distaste for following due 'bureaucratic procedures' necessary to gain access to funding:

> In the Netherlands I have developed an extreme distaste for writing down plans for nothing, even though my score at the KWF was over 80 per cent over 30 years of cancer research. So I have written a concise plan to Mr Broesterhuizen and I shall add a first draft of our planned home page to be called 'The Indonesian Hereditary Disease Registry'. [22 February 2002, to PF; transl. MS-F]

Having been successful and having valuable resources on offer, Dr X preferred to disregard due procedures:

> Now I do hope that you won't be sending me a pile of forms, for then I will get up on my hind legs. I am so happy to have been able to avoid them here in Indonesia over the last three years. The precise contents of the plans will come later, me thinks [*sic*], and I shall welcome any suggestion from the Netherlands. We are prepared to take on anything as long as we can establish a beginning of cooperation. There are hundreds here raring to go but can't find a job and if they do they have no money to conduct experiments. A sad situation here, apart from with me at Sanbe Farma, fortunately. [22 February 2002, to PF; transl. MS-F]

Dr X seemed to know how to pull the strings in the Indonesian bureaucracy and understands how to approach the right person in the right institutions:

> Furthermore, we are setting up one or more 'field' teams that carry out expeditions, such as to Papua and the inlands of Sulawesi, Kalimantan and Sumatra. To approach small ethnic groups permits are required from a specially designated Ministry. All contact with the Government from here: NOT OTHER WAY ROUND PLEASE. So Louk, Emil and others, please, do not play the big guy in Jakarta. It is better if we, i.e., Pak Jahja, my boss at Sanbe[32] and 'Kus', the Rector of ITB, present ourselves as 'small boys' with an offer from the Netherlands. We know the 'right price' to pay (you don't, which would end in a right disaster) named KKN [corruption, collusion and nepotism][33] (try to translate that in proper Dutch). [23 February 2002; transl. MS-F]

The special nature of the Indonesian bureaucracy apparently requires special treatment.

But when about four months later official enthusiasm subsided, Dr X started to question the use of continuing the discussion with geneticists in universities. [24 June 2002, to CB] When 'asked what he really wanted' [24 June 2002, to CB], Dr X set his aims less high, stating that he did not really want to do much more than what is common practice in the Netherlands already: collecting materials from families with genetic deviations to perform routine molecular diagnostics. Surmising that the new Dutch infrastructure was 'rather enormous', Dr X also considered trying to find susceptibility genes indicative for complex genetic diseases. But it was not really his field – making a contribution to the Human Genetic Diversity Project had been his main scientific dream [24 June 2002, to CB].

It was Dr X's institutional position that enabled him to mobilise contacts from Southeast Asia, apply for funding in the Netherlands, using his contacts in Dutch universities and government organs, employ cheap, highly educated labour in Indonesia, and pull strings in the Indonesian bureaucracy. Dr X's privileged position cannot be understood as it was generated, in part, thanks to the institutional structures and relations of inequality prevailing under oppressive regimes in the Dutch colonial and republican postcolonial eras.

Pulling strings and mobilising support

It was not only the Dutch–Indonesian colonial history, but also global developments in biotechnology that have produced the institutional conditions for an entrepreneur cum scientists (cum poet) to mobilise potentially fruitful partnerships in biobanking. Vis-à-vis his colleagues, Dr X put up a show of feeling at home and of being respected far away from home in Indonesia, while making sure to communicate the impression that he could move mountains at will. Both showing off how he rubbed shoulders with the world's greatest and giving the impression that he (and the other greats) could take on the challenge of finding the key to Asian genetic diversity were not only an expression of his vanity, but, more importantly, conditioned by the institutions that he was in a position to mobilise.

Vis-à-vis official organisations, he could not take the same stance – showing at least some respect for the rules he believed it was necessary to get past. As in any bureaucracy, there are ways of manipulating bureaucratic administration, such as by searching for alternative readings of the rules or by simply playing on the sensitivities of a functionary. The bureaucrat, in turn, can learn how to deal with troublesome cases and may take steps in order to improve procedures. In the case of Dr X, his unwillingness to follow official procedures in the Netherlands was accompanied by pulling strings among his contacts in the Netherlands. In the end it did not yield any funding, which was largely reserved for the Genome Initiative in the Netherlands.

The Indonesian bureaucracy, however, has no effective standard procedures.

Individuals in the bureaucracy have control over resources that they use to build up internal and external relations of reciprocity. The prices of 'their' resources are variable, depending on the relation with and the resources of the clientele. In dealing with such a bureaucracy the client needs tact and humility. Dutch big shots insisting on their rights could endanger existing deals between the authorities and Sanbe Farma.

It is clear that in the Netherlands there was more funding available than in Indonesia (despite Dr X positive experience at Sanbe Farma). High-quality researchers in Indonesia were cheaper than they were in the Netherlands, and were highly motivated and disciplined. In Indonesia, healthcare institutions can hardly cover the needs of the population, but they were mobilised to cooperate in projects:

> Now that we have involved the Puskesmas [Community health centres] (sixty-eight in total) of Bandung, things are really starting to get cracking and the aim is to involve the local healthcare organisation of the entire country in an enormous national programme, for which there are more then enough means available. [21 February 2002, to FH; transl. MS-F]

Such determined optimism leads one to believe that Dr X either had no clue about the existence of research regulations or that he ignored them.[34]

It should be noted that the NCEHR was set up after Dr X's e-discussion on the Indonesian/Dutch Genome initiative took place. That may be why not a word has been dedicated to informed consent procedures, the explanation of the research procedures, and its medical and financial purposes to laypeople. No consideration was given to the political consequences for the individuals involved, their families and the ethnic communities. But how will patient confidentiality with regard to the identity and health records be maintained if, according to government regulations, 'Requests for help in gene mutation searches will be given only if the pedigree is as complete as possible, if relevant medical records are sufficient?'[35] No effort is made to elaborate on how those who are found genetically susceptible to, for instance, familial cancer and mental retardation will have to deal with their condition, and what precautions and what measures should be taken. These questions are all the more urgent considering that historically in Indonesian society there has been a tendency to regard patients with mental illness as a threat to those around them rather than as sick people in need of support and care.

Conclusion

In this article, the author has told Dr X's story from various points of view. Dr X as a scientist, an entrepreneur and an enlightened thinker. Indeed, his expertise, his contacts, his altruism, his enlightened thinking, his exploitation of cheap labour, his Orientalist attitude and his altruism all played a role. The author hopes to have shown, however, that in the end, the institutional set-up in which Dr X operated was of greater importance than his motivations for setting up a genetic registry. It was the presence of historical and institutional ties between the Netherlands and Indonesia, the relative difference in the prices of educated labour, his access to funding, and disjointed welfare and legal institutions that made Indonesia the location in which he could realise his dream: a country that recently has experienced great political turmoil and economic calamity.

Without the awareness of biological citizenship in modern biosocieties and the important role played by their genetic make-up in genetic registry, and without adequate healthcare provision, it is tempting for impoverished families to join a genetic sampling project for little in return. The attraction to genetic samplers of Indonesian people as biological resources does not just lie in the unique and valuable nature of their DNA, but in the relative ease with which research permits are received, bureaucratic procedures can be avoided, the low price of schooled labour, and the relative lack of resistance that vulnerable (illiterate, uneducated, or poor) people display to becoming targets of research. The fact that Indonesia does not have a genomics initiative, but has insufficient public health resources and little bioethical protection means that the country formed an ideal research target. This situation makes it almost inevitable that Indonesia will be forced to devise bioethical review and regulation and spread 'genetic awareness'. The effect of such biological citizenship in Indonesia may have far-reaching implications for its ethnic strife, socio-economic inequality, and political nationalism.

The perspectives of progress and exploitation were used heuristically to highlight the ways in which transnational interstices or gaps between regulatory, educational and welfare provisions in modern biosocieties and pre-biosocieties forces the latter to review their political and institutional infrastructure. However, modern biosocieties have the advantage of being able to manoeuvre across borders without many problems, while developing countries are often limited by the relative number of opportunities to move transnationally, and by ineffective political and legal regulatory institutions at home. Thus, Dr X manipulated the differences between national institutions, combined his networks and created plans that took advantage of both of the worlds he occupied: Dutch education and Indonesian experience, Dutch academic and political networks and funding and Indonesian companies, expertise and resources. Indeed, he took advantage of the bioethical gap, the socio-economic gap, and the regulatory gap between Dutch and Indonesian institutions: the transnational interstices and differences of available knowledge, human biological resources, healthcare facilities, financial resources, and expertise in the Netherlands and Indonesia.

Notes

1 Even though the name of Dr X is retrievable, I do not use it because this article is about diverging social and political institutions among which Dr X operates, not about Dr X *per se*.

2 I do not believe it necessary to draw attention to persons that are only loosely connected to the Dr X case.
3 PT Sanbe Farma employs 5000 people, has a turnover of US$50 million (50 per cent growth in 2001), 30 sales offices, 600 medical representatives, 2000 people active in the distribution system and over 40,000 client doctors (21 February 2002).
4 Dr Emil Broesterhuizen in 2000 became deputy director of Research and Science Policy at the Ministry of Education, Culture and Science, which he combined with the leadership of the Department of General Issues of Science Policy and International Co-operation of the board of OWB until 2001.
5 For mDNA, see the work by Sangkot Marzuki: see PubMed under <smarzuki>).
6 In his August 2002 essay, he has raised the number to 600 (cf. www.sanbe-farma. com/biotech.html accessed 8 February 2006).
7 At the time, the home page had not been set up as they only had about ten families.
8 22 February 2002.
9 To alleviate the shortage, Dr X's Division of Biotech and Research started to produce reagents locally such as Taq polymerase for PCR reactions and DNA isolation kits.
10 Koninklijk Instituut voor de Tropen (KIT).
11 Memoranda of Understanding are in place with two Universities of Bandung, one Institute in Jakarta and one University in Medan, with firmly established links in, Semarang, Yogyakarta, Palembang, Surabaya, Makassar and Bali/Lombok (cf. http://www.ihdreg.com/mainpage.htm).
12 Dr X's slideshow 'The Indonesian Hereditary Disease Registry' on http://www. ihdreg.com/mainpage.htm (accessed 8 February 2006).
13 The Institute has established international reputations in such arenas as mitochondrial diseases and the study of haemoglobinopathies, and serves as Indonesia's national resource centre for genetic testing (http://www.eijkman-institute.org).
14 1 August 2002 (http://www.ihdreg.com/mainpage.htm).
15 For a long list of declarations against the HGDP, see Macer 1997: note 6.
16 Sites in the DNA sequence where individuals differ at a single DNA base are called single nucleotide polymorphisms (SNPs). A haplotype is a pattern of SNPs on a block of inherited sets of nearby SNPs on the same chromosome.
17 To locate such genes, researchers perform linkage analysis on families, which requires 300–500 highly informative genetic markers spanning the entire human genome.
18 For an analysis of race as a social category see Wade 2002.
19 Two years later there is still no evidence that ethnic labels on the DNA of samples collected from his families are in use (cf. http://www.ihdreg.com/mainpage.htm [accessed 8 February 2006]).
20 See www.gis.a-star.edu.sg/homepage/gismediapress.jsp?pid=33
21 Sultana Muhammad Hussein obtained his PhD entitled 'Fragile X Chromosomes in the Indonesian Population form the Faculty of Medicine' from the University of New South Wales, Sydney (December 1998).
22 Due to its closed nature, the population of the village of Giethoorn is believed to deviate genetically from the rest of the Dutch population.
23 ADB (Asian Development Outlook) GDP of Indonesia: 1997: 3,2; 1998: -14,7; 1999: -0,7; 2000: 3,3; 2001: 1,9; 2002: 2.2 (Bergstrom 2004: 132).
24 UNDP-Human Development Report 2001. See: http://hdr.undp.org/reports/ global/2001/en/; Bergstrom 2004: 134; 142.
25 Brief report of ICMR–WHO Inter-country meeting on Ethics in Health Research, held from 9–13 September, 2002, at Hotel Le Meridien, New Delhi, India. See: icmr.nic.in/bioethics/ ICMR–WHO%20Intercountry%20meeting.pdf
26 Also take into account that the adult illiteracy rate is 13.7 per cent and that only 13

per 1000 Indonesians owned a PC in 1999 (WDP Report 1999/2000, Bergstrom 2004: 322).

27 Government Regulation of the Republic of Indonesia No. 39 of 1995 concerning health research and development. Dated 14 November 1995 Indonesia. Decree No. 574/MENKES/SK/IV/2000 of 10 April 2000 of the Minister of Health (MOH) on health development policy toward a healthy Indonesia 2010. http://www.who.int/idhl/

28 Established by Decree of MOH No 1334 / 2002.

29 Bioethics related activities in Indonesia 2005: MOH No 1334 / 2002.

30 Two MOUs were signed by the Dutch and Indonesian ministers in support of the Scientific Programme Indonesia-Netherlands (SPIN) coordinated by The Royal Netherlands Academy of Arts and Sciences (KNAW). Since 1997, infectious diseases and biotechnology constitute two of the six focus points of research (Stapel 2003).

31 Dr Edison Liu, Executive Director of the Genomics Institute of Singapore (GIS).

32 Dr Jahja Santoso, Apt., President Director of Sanbe Farma.

33 Most Indonesians are agreed that political 'reformasi' since 1998 and the fall of Suharto have failed to check corruption, commonly referred to as KKN. The government of President Susilo Bambang Yudhoyono has made the fight against KKN its chief priority, but the results so far have been disappointing.

34 The Government Regulation of the Republic of Indonesia No. 39 of 1995 concerning health research and development was established on 14 November 1995.

35 Dr X, 1 August 2002, http://www.ihdreg.com/mainpage.htm.

References

Abdullah, Irwan, and Hüsken, Frans (2003) Social security research in a time of crisis. In: Stapel J. (Ed.) *Scientific Programme Indonesia–Netherlands (SPIN)*. Amsterdam: KNAW, 17–26.

Avise, John C. (2001) *The Genetic Gods. Evolution and Belief in Human Affairs*. Cambridge, MA: Harvard University Press.

Bergstrom, P. (Ed.) (2004) *Ethics in Asia-Pacific*. Bangkok: UNESCO.

Busby, Helen, and Martin, Paul (2006) Biobanks, national identity and imagined communities: the case of UK Biobank. *Science as Culture* (forthcoming).

Carlson, Rick J., and Stimeling, Gary (2002) *The Terrible Gift. The Brave New World of Genetic Medicine*. New York: Public Affairs.

Cavalli-Sforza, L.L., and Cavalli-Sforza, F. (1995) *The Great Human Diasporas. The History of Diversity and Evolution*. Cambridge, MA: Perseus Books.

Cavalli-Sforza, L.L., Wilson, A.C., Cantor, C.R., Cook-Deegan, R.M., and King, M.C. (1991) Call for a worldwide survey of human genetic diversity: a vanishing opportunity for the human Genome Project, *Genomics* 11: 490–1.

Desowitz, Robert S. (2002) *Federal Bodysnatchers and the New Guinea Virus*. New York: Norton & Company.

Gordimer, Nadine (2002) A New Racism, World Watch, July/August 2002, http://www.genetics-and-society.org/resources/items/200207_worldwatch_gordimer.html

Harris, John (1998) *Clones, Genes and Immortality*. Oxford: Oxford University Press, Chapter 7.

Harry, Debra, Howard, Stephanie, and Shelton, Brett Lee (2000) Indigenous People,

Genes and Genetics, Indigenous Peoples Council on Biocolonialism (IPCB), June 2000; cf the IPCB website available at http://www.ipcb.org/

Heath, Deborah, Rapp, Rayna, and Taussig, Karen-Sue (2003) Genetic citizenship. In: Nugent, David, and Vincent, Joan (Eds) *Companion to the Handbook of Political Anthropology*. London: Blackwell.

Knoppers, Bartha (2003) *Populations and Genetics: Legal and Socio-Ethical Perspectives*. Leiden: Martinus Nijhoff.

Lewontin, Richard (2001) *It Ain't Necessarily So. The Dream of the Human Genome and Other Illusions*. New York: New York Review Books.

Lock, Margaret (1994) Interrogating the human diversity genome project, *Social Science & Medicine* 39: 603–6.

Macer, D.R. (1997) Bioethics and genetic diversity from the perspective of UNESCO and non-governmental organization. In: Knoppers, B.M. *et al*. (Eds) *Human DNA: Law and Policy. International and Comparative Perspectives*. The Hague: Kluwer Law International.

Marshall, Eliot (1997) Gene prospecting in remote populations, *Science* 278(5338): 565.

Multatuli (1969) [1875] *Max Havelaar*. Rotterdam: A.D. Donker.

Normile, Dennis (2004) Genetic diversity: consortium hopes to map human history, *Asia Science* 306(5702): 1667.

Peterson, Anna L. (2001) *Being Human. Ethics, Environment, and Our Place in the World*. Berkeley, CA: California University Press.

Pilnick, Alison (2002) *Genetics and Society: An Introduction*. Buckingham: Open University Press.

Rabinow, Paul (1992) Artificiality and enlightenment: from sociobiology to bio-sociality. In: Crary, Jonathan, and Kwinzer, Sanford (Eds) *Incorporations*. New York: Zone, 234–52.

Rawls, John (1973) *A Theory of Justice*. Oxford: Oxford University Press.

Reardon, Jenny (2001) The Human Genome Diversity Project: a case study in co-production, *Social Studies of Science* 31(3): 357–88.

Reardon, Jenny (2004) *Race to the Finish: Identity and Governance in an Age of Genomics*. Princeton, NJ: Princeton University Press.

Rifkin, Jeremy (1998) *The Biotech Century*. London: Orion Books.

Rose, Nikolas, and Novas, Carlos (2003) Biological citizenship. In: Ong, Aihwa, and Collier, Stephen (Eds) *Global Anthropology*. London: Blackwell.

Said, Edward W. (1979) *Orientalism*. New York: Vintage Books.

Sándor, Judit (2003) *Society and Genetic Information. Codes and Laws in the Genetic Era*. Budapest/New York: Central European University Press/CPS Books.

Savalescu, Julian (2001) Procreative beneficence: why we should select the best children, *Bioethics* 2001: 415–26.

Sjamsuhidajat, R. (2004) Ethics in medical education and medical practice: Indonesia. In: Kasturiaratchi, Nimal, Lie, Reidar, and Seeberg, Jens (Eds.) *Health Ethics in Six SEAR Countries. Edited Health Ethics in Southeast Asia Volume 1*. New Delhi: World Health Organization Regional Office for South-East Asia.

Slamet, Lucky S., and Elengold, Mark. A. (2002) Regulating Biotechnology Products, available at www.who.int/medicines/library/qsm/icdra02/pdf/10.pdf-link.pdf

Stapel, Johan (ed.) (2003) Scientific Programme Indonesia–Netherlands (SPIN). Amsterdam: KNAW. Available at: www.knaw.nl/publicaties/pdf/20031027.pdf

Stock, Gregory (2003) *Redesigning Humans*. London: Profile Books.
Sutrop, Margit (Ed.) (2004) *Trames. Journal of the Humanities and Social Sciences* 8 (1/2): Special issue: 'Human genetic databases: ethical, legal and social issues'.
Sykes, Bill (2002) *The Seven Daughters of Eve*. London: Corgi Books.
Wade, Peter (2002) *Race, Nature and Culture. An Anthropological Perspective*. London: Pluto Press.

7 India's preparedness in tackling biopiracy and biobanking

Still miles to go

Nandini K. Kumar

Introduction

According to official regulations, in India the analysis of DNA samples has to be carried out by Indian scientists or laboratories and no sample can be sent out without following the guidelines of the Government of India. Despite steps to ensure protection of human rights and sensitivities of the participating individuals and populations involved in research there have been many violations, not to speak of intellectual property rights (IPRs) being overlooked.

The Department of Biotechnology (DBT) proposes to study human genome diversity in India and envisages a national repository in collaboration with the Anthropological Survey of India and other related agencies. The ethical guidelines of both the Indian Council of Medical Research (ICMR) and the DBT in the area of genetics include some aspects of biobanking, which will come out as a single code when they will be legislated shortly in 2007. According to these guidelines for obtaining samples, relevant issues such as informed consent, the collection of samples and their purpose, access to these resources, ownership rights of DNA, and quality control of the laboratories where samples are tested should all be appropriately documented in the research proposal for review by the Institutional Ethics Committees. Despite these safeguards, which are not yet mandatory, India, with 4635 castes, 3000 communities and 28,000 endogamous groups, risks ethnic disharmony through information generated by studies on genetic susceptibility. Moreover, the potential of obtaining valuable information which may lead to academic laurels or commercial benefits is leading to increasing instances of biopiracy. There is a lack of adherence to regulations existent in the country due to the loose nature of these regulations, which hardly lead to any punitive measures being implemented. The immenseness of the country with its teaming millions poses problems for policing in this regard.

The main demands envisaged for obtaining samples are those for studies on anthropology, pharmacogenetics, disease susceptibility, and forensic purposes. There is also an urgency for the preservation of samples of ethnic groups who may be about to become extinct. While on the one hand it would be the right time to consider the ethical, social, and legal issues involved in biobanking, on the other hand strong steps will have to be taken to stem biopiracy.

Genetic sampling and its regulatory complexity

Being a biodiversity-rich country, India attracts global interest in the study of genomic diversity of anthropologically well-defined populations including tribal ones. This not only has global evolutionary implications but also has an application in pharmacogenetics for personalised medicine.[1,2] India's strength in the area of drug development further complicates the situation, especially as Indian biobanks, as national repositories of biological samples, give rise to issues such as privacy protection and confidentiality. When international collaborative studies are undertaken, some main concerns are exploitation, stigmatisation, and benefit sharing, especially when a gene- based product becomes commercialised. Biopiracy is even more worrisome if the developing country has appropriate ethical guidelines and regulations in place, but lacks proper implementation mechanisms to monitor misuse.

Against such a background, setting up a biobank in India is fraught with problems. It not only involves huge investment but also requires a well-thought-out plan for its governance, taking into account the ethical and legal aspects relevant to India. There are efforts on the part of the government and private concerns to create biobanks for different purposes. While agencies of central government such as the Anthropological Survey of India and the Ministry of Culture are interested in preserving the biological samples or DNA of ethnically divergent groups, especially those on the verge of extinction, private firms have been engaged mostly in preserving stem cell-rich cord blood, mainly for transplantation or regenerative therapeutic purposes. Already, the Centre for Cellular and Molecular Biology, Hyderabad has collected 9000 samples from 130 endogamous populations from India.[3] A DNA Profiling Advisory Committee (D-PAC), set up by the Union Cabinet under the chairmanship of the Prime Minister, has been formed. It will not only suggest how to create a national DNA profiling facility but will also recommend draft legislation to start DNA profiling in India.[4] Reliance Life Sciences and Life Cell (the Asian Cryo Cell company in collaboration with Cryo Cell International Inc.) are two private concerns which have obtained license from the Drugs Controller General of India for setting up a cord blood bank. Reliance Life Sciences has collected about 3000 cord blood samples, while Life Cell is fairly new in the area. The Institute of Immunohaematology, an ICMR Institute, was initially funded by the DBT to set up a repository for studies on heamatological disorders. But beyond studies on cryopreservation of blood cells, further steps have not been taken due to various reasons involving expansion of such a facility. There are other premier educational institutions, such as the All India Institute of Medical Sciences, New Delhi, and Christian Medical College, Vellore, and some private institutions and organisations planning to set up biobanks for basic and therapeutic research. Given India's varied patient population as subjects for drug trials, and the large number of clinical research organisations setting up shop in the country, there is a great likelihood of biological material being sent out of the country due to its commercial potential in the name of academic studies on pharmacogenetics. There are guidelines and regulations extant in India with regard to use of biological materials to prevent misuse. The anonymity of samples can safeguard privacy and confidentiality of the donor or the community. But these very aspects could negate the benefits reaching the participants if there is a positive outcome of pharmacogenetics research in diagnostic or therapeutic terms or if commercialisation later develops.

Guidelines, rules and regulations

In India, the analysis of DNA samples will be carried out by Indian scientists and laboratories and no sample can be sent out without following the guidelines of the Government of India in this regard. The Ministry of Health and Family Welfare formulated guidelines for transfer of biological material in 1997[5] and all international collaborative studies have to comply with these governmental directives. International collaboration requiring biological samples to be taken out of the country or their exchange is possible only if a project proposal with a clear and well-documented memorandum of understanding (MOU) has been approved by the Institutional Ethical Committee and the Health Ministry's Screening Committee (HMSC). In the event of failure of such agreement, the guidelines of the country (India) prevail. When a tribal or ethnic group is targeted for research, permission from the Tribal Welfare Commissioner of the relevant state has to be obtained before approaching the group.

The guidelines or regulating factors that apply to biological samples and the ethical issues that could arise in related instances are as follows:

- Ethical Principles – General and Specific on Transplantation chapter in the ICMR's revised ethical guidelines of 2000.[6]
  ICMR guidelines state that for obtaining samples relevant issues such as consent, collection of samples, access to these samples, and the purpose for which they are accessed, ownership rights of the DNA, and quality control of the laboratories where samples would be tested should all be appropriately documented in the research proposal for review by the Institutional Ethics Committees. India, with 4635 castes, 3000 communities and 28,000 endogamous groups, is fertile ground for riots among ethnic groups for smaller reasons. In such a situation, generating information through genetic studies which could affect cultural practices of a community, a major concern would be the possibility of creating ethnic disharmony.
- Health Ministry's Screening Committee
  For more than two decades all foreign-funded projects have been required to be screened by this committee, and the revised ethical guidelines of the ICMR[7] were released in 2000. Still, there have been instances of violations of these requirements. Not to speak of IPR issues, concerns for human

rights and protection of sensitivities of the participating individuals and populations have also not been considered when Indian researchers intentionally or through ignorance of regulatory requirements have allowed or taken samples out of the country.

- The Government of India's order on 'Exchange of Human Biological Material for Biomedical Research'
  This order, by the Ministry of Health and Family Welfare for transfer of biological materials, was passed in 1997 and all international collaborative studies have to comply with the government directives mentioned in the order.[8]
- The designation of repositories under section 39 of Draft Rules Biodiversity Act states:[9]

1 The Central Government in consultation with the Authority may designate for maintaining a database on Biological resources national premier institutions, involved with various categories of biological diversity as repositories; such as Botanical Survey of India, Zoological Survey of India, National Bureau of Plant Genetic Resources, National Bureau of Animal Genetic Resources, National Bureau of Fish Genetic Resources, Institute of Microbial Technology, National Institute of Oceanography, etc.

2 The repositories designated in sub-rule (1) shall take steps for safe custody of biological material accessed including DNA fingerprints, if necessary.

3 The repositories shall maintain an electronic database of the collections of the biological resources available with them and submit a periodical report to the Authority and the Central Government.

4 The repositories shall provide such information as may be asked for by the Authority, State Biodiversity Board, and the Biodiversity Management Committees for successful implementation of the Act.

5 The Central Government may consider providing necessary financial assistance to designated repositories for the upkeep of the specimens and their maintenance.

At present, India is in the process of drafting guidelines for biobanking. The main demands for samples envisaged are for studies on anthropology, pharmacogenetics, pharmacogenomics, disease susceptibility, and forensic purposes. There is national urgency to preserve for posterity biological material of some ethnic groups, who are rapidly becoming extinct. This would be the right time to consider how to ensure informed consent for donating samples to the biobank and under what circumstances the results of the research should be communicated. ICMR guidelines state that results should be conveyed for only those diseases for which there is some intervention possible. However, keeping in view possible future trends, which may bring relief of disease symptoms much earlier than thought at present, one may have to reconsider these guidelines about the communication of results. Linkage of genetic databases to the healthcare system may not be a far-off possibility. Except in conditions which may cause public harm – for instance, employing a pilot affected by sickle cell anaemia – linkage of genetic database to a database of third parties should be considered on a case-to-case basis.

The acceptability of a community to donate samples assumes great importance in a land with a potential for negative political usage because many ethnic groups form the voting blocks of local politicians who may object to particular objectives of research. Therefore, without prior counselling at an individual level and consultations at community level, samples should not be collected. The ownership of a sample of a pre-defined community spread over different geographical areas but with different socio-political practices and cultural traditions creates another ethical concern to be resolved, especially when it comes to benefit sharing. For instance, members of a named tribe of the south of India in one particular state may have migrated to other southern states long ago and, therefore, now they may be genetically different through mutations effected by the newer environment. If the biological samples obtained from the tribe in one particular area contributes to the development of a diagnostic or therapeutic product, and this group gets commercial benefit from it, then the members of the same tribe settled in other southern states could also claim their share of benefit. It would be an ethical debate whether this could be allowed. Depending on political and social interest and the stand of commercial organisations in this matter, it could be become a major issue. This question arose when the Kani tribe of Kerala, a southern state, were given a substantial share of benefits because they held the knowledge of a medicinal plant used by them to alleviate stress. The question raised was whether the Kani tribes residing in other neighbouring states were also to be given benefit. The remuneration/share being sought by the participants from private institutions or pharmaceutical companies should be fixed after appropriate consultation with the authorities or even at the national level.

There should be a national repository for tribal studies, as many tribes are unapproachable and have become scant in number. It would also safeguard the interest of the tribes and protect them from possible exploitation by commercial concerns through investigators and companies such as IuVita.[10] Any submission for obtaining samples should not be without the approval of an ethics committee, a procedure that will become mandatory once the ethical guidelines are legislated. However, for drug trials involving pharmacogenetics, it is mandatory to get the ethics committee's approval as per the revised Schedule Y of the Drugs & Cosmetics Act (1940) amendment February 20, 2006. In all matters, the role of the Government is significant.

Research on genetics, genomics, and plant-based medicine is threatening to cause a technical and resources divide between the developed and the developing countries similar to the 10/90 gap.[11] This North-South divide is relevant here: if the North is technically superior, then the South is richer in resources, both in flora and fauna. However, in most of these genetic resource-rich countries their poor regulatory mechanism leads to exploitation of these resources. Biopiracy in the present scenario is even more worrisome if the developing country has ethical and regulatory mechanism in place to prevent such exploitation but does not have an effective monitoring or implementing mechanism.

India is one of the twelve biodiversity-rich countries in the world. Just as the fever of revelation of the human genome was catching on, India proceeded to safeguard its interests by initiating the formulation of guidelines to prevent misuse of biological samples. But the potential of obtaining valuable information, which may lead to academic laurels or commercial benefits, is leading to increasing instances of biopiracy. Of late, more and more instances are coming to light as these guidelines are flouted by researchers who are associated with foreign investigators funded by agencies, which in their own countries are required to adhere strictly to high ethical standards concerning human research. For instance, samples from Indian tribes were taken abroad for genetic analysis without permission being obtained from the local authorities concerned. The study was published in the *American Journal of Human Genetics*,[12] and listed eighteen authors from seven institutions in six countries. The National Institutes of Health (USA), the European Union and Estonian agencies funded this study. A science journalist brought this matter to the attention of ICMR. It was then noted that not a single Indian institution or author was acknowledged, and it is not clear who gave permission to collect the samples. It is also mentioned in the article that informed consent was taken. When ICMR further enquired into the issue it was communicated by the EU Commission that the samples had been collected twenty-five years previously. Considering that the concept of informed consent is still not understood by Indian patients/volunteers, what the condition could have been twenty-five years ago is quite unimaginable. At the same time, it also raises the issue of how samples stored with a foreign collaborator should be handled with regard to obtaining informed consent years later from the participants from the host country. In anthropological studies, anonymising such samples for want of consent will carry no meaning.[13]

When it comes to using biological samples, fresh or stored, especially for genetic research, informed consent of individual, family, community, and even government (in case of international collaboration), will be required before the initiation of such projects. Therefore, great care is necessary when releasing the results to the media or publishing such results, because a potential psycho-social harm exists in the form of anxiety and depression or damage to familial or community relationships. For instance, marriage prospects of girls of a village could be affected if damaging results appear in the media based on a report of a genetic study involving that community.[14] Without ensuring that these requirements are complied with, the publication of such studies should be discouraged. In this age of pharmacogenomics, this would prevent adverse socio-economic consequences and the unnecessary generation of benefit sharing and caste-based political issues relating to IPR and patent rights of biodiversity-rich developing countries if the stored samples lead to commercialisation of leads obtained from testing those samples.

As described earlier in this article, when research involves the tribes of India, which are considered to be vulnerable populations, permission of the Tribal Welfare Commissioner of the concerned state is required. However, in practice, not only are their samples being carted away surreptitiously, their stored samples are also reused without obtaining fresh consent. Even the Indian researchers in developed countries and within the country involved in the area are ignorant that India has a regulatory mechanism in place to prevent such collection.[15] It is difficult for the government to police such activities. However, a published article, the end result of endeavours for glory, can pin down such violations at international level. Could publication of anthropological studies in foreign journals, which have not screened the ethical aspects of how the samples were obtained, possibly lead to the social stigmatisation and discrimination in employment, health, and general insurance of the participants once the results get known in their country? Moreover, in international collaborative studies, due adherence to Indian regulations should be given and Indian contribution should be acknowledged. The sad fact is that even after informing the journal editor of the *American Journal on Human Genetics* of the bioethical issue of informed consent in relation to the study on which the article on the genetic heritage of Indian tribes was based, no action was taken and no Indian investigator or institution was acknowledged. When this was brought to the attention of the concerned ethics committees of the foreign funding agencies, some enquiry was made but not pursued to a satisfactory conclusion. Later, some more articles[16,17] on the genetic make-up of Indian tribes were published but this time with the name of an Indian investigator and institution included. But it appears that approval from concerned authorities was not obtained for sending the samples abroad for the study.

Within India, clinical research organisations are now engaged in creating websites to attract commercial offers for their databases on clinical profiles and for biological samples of diverse patient populations under their study. This unscrupulous practice, unless detected, will deprive these unsuspecting patients of any commercial benefits that may accrue through patents obtained by pharmaceutical companies in a surreptitious manner. The onus to be vigilant about this in collaborative studies falls on the editorial boards of journals and ethics committees of the developed world's partner institutions, as ethics committees functioning in developing countries are still in an early stage of development.

Conclusion

India has a system to check exploitation of human participants involved in research but this will only be effective if the researchers, ethics committee members and editorial board members are aware that malpractice should be adequately punished, and the political will is there to implement the system, so that India can benefit from her rich biodiverse heritage. Biopiracy should be tackled vigorously by bringing into one fold the various stakeholders as advisors on a national committee, and by enforcing punitive measures to check misuse and unethical commercial advertisements. Biobanking will become more meaningful with appropriate governance principles, but India has still a long way to go in formulating guidelines, including those for the private sector. Indian scientists and populations should be given due recognition in international collaborative studies, which should be well screened and regulated.

Notes

1　O'Kane, D.J. and Moyer, T.P. (2003) Pharmacogenomics and reducing the frequency of adverse drug events, Editorial in *Pharmacogenomics*, 41: 1–4.
2　Daar A.S. and Singer, P.A. (2005) Opinion, pharmacogenetics and geographical ancestry: implications for drug development and global health, *Nature Reviews Genetics*, 6: 241.
3　Available at: http://www.rediff.com/money/2005/jun/13gene.htm (last accessed on 10 January 2007).
4　Available at: http://www.deccanherald.com/deccanherald/dec14/n9.asp (last accessed on 10 January 2007).
5　See Government of India. Exchange of Human Biological Material for Biomedical Research. New Delhi: 1997, available at www.icmr.nic.in
6　Indian Council of Medical Research. 'Ethical Guidelines for Biomedical Research on Human Subjects', 2000, New Delhi.
7　These guidelines have been acknowledged by USA as equivalent to their own guidelines for protection of human participants in research.
8　Available at www.icmr.nic.in
9　Draft Rules Biodiversity Act (2002) has been prepared under the Ministry of Environment and Forests.
10　The company IuVita Biobank is one of many which has been advertising the availability of biological samples of the Indian population on the web. IuVita, a private company, is planning a biospecimen repository of the Indian patient population.
11　The 10/90 gap refers to the utilisation of 90 per cent of resources on R & D for relief of 10 per cent of the population of the developed countries, the North. Similarly, 90 per cent of disease burden is in developing countries but the populations of these countries are used for finding treatment options for the 10 per cent residing in countries of the North.
12　Kivisild, T., Rootsi, S. Metspalu, M., *et al.* (2003). The genetic heritage of the earliest settlers persists both in indian tribal and caste populations, *American Journal of Human Geneticts*, 72: 313–32.
13　The samples were taken from India by foreign investigators (most likely by the Indian investigators of the foreign institutions in the research group). It was conveyed by EU Commission that since the samples were collected and kept in a foreign repository for all these years, it would not have been possible to obtain informed consent. If that is so, then why was it stated in the article that informed consent had been taken? Was it due to the insistence by the journal about this inclusion without which it would not have been possible to publish the paper? The latter view was the answer given by a bioethicist during a recent Indo-US workshop on bioethics, which was instrumental in bringing the subject to the notice of the EU.
14　This has already happened as a result of reports published on high rates of HIV in India in some geographical locations. In India, caste wars happen for more trivial reasons, such as disputes over sharing well water, a common news item in the press. Any information associated with the genetic make-up of a caste or tribe would cause havoc.

15 There have been publications in peer-reviewed journals where the investigators from the North have obtained samples from Indian tribal populations without permission from local relevant authorities and have used samples stored for some 20–25 years. What must have been the nature of informed consent taken then as recorded in the published articles?

16 Cordaux, R., Saha, N., Bentley, G.R., *et al.* (2003) Mitochondrial DNA analysis reveals diverse histories of tribal populations from India, *European Journal of Human Genetics*, 11: 253–264.

17 Cordaux, R., Aunger, R., Bentley, G., *et al.* (2004) Independent origins of Indian caste and tribal paternal lineages, *Current Biology*, 14: 231–235.

8 The Indian genomic biobank initiative and emerging bioethical issues

A community-based perspective

Prasanna Kumar Patra and
Margaret Sleeboom-Faulkner

Introduction

India, a still developing industrialised country, is debating the establishment of a national biobank or national repository (NR), an initiative open to ethical, legal, and social scrutiny. This biobank initiative started with the idea to provide a centralised repository for the randomly and unsystematically collected and stored biospecimens and health data at various research centres, including universities, public and private hospitals, and research institutes. The mapping and enumeration of such banks is difficult to comprehend and describe in brief. The biological specimens and health data were generally gathered through two processes: first, through genetic, epidemiological, and carrier-screening programmes in the community; and second, from individuals and patients in hospital situations. At a community level, it is the population geneticist, local hospitals, and physical anthropologists who have been involved in data generation; at a hospital level, it is individual patients who provide data through undergoing testing or tissue donations.

India's proposed national biobank initiative will be under scrutiny because of enormous social, economic, and healthcare disparity and a lack of well-defined and stringent regulatory mechanisms for the conduct of biomedical research. In this paper an attempt is made to stimulate discussion on common bioethical issues, such as informed consent, confidentiality, benefit sharing, and public trust. These issues are considered to be vital in the field of genomics in India from a community perspective based on anthropological studies.

India's biobank initiative

The endeavour of establishing a permanent NR for human genetic resources and data in India is a by-product of developments in other fields. The last one and a half decades have seen India as an emerging economic power, after liberalisation of its economy in early 1990s, maintaining a high and steady growth rate at about 7–8 per cent of its GDP. It acquired most of its leverages from service sectors such as information technology (IT), in which it boasts itself to be a world leader. In fact, India's success in information technology is such that its computer software and services industry grew from about US$500 million to more than US$10 billion in exports over the last decade.[1] This has spawned a burgeoning bioinformatics industry.[2] The profile of Indian biotech companies is undergoing change, and the companies have come to play an important part in the innovation of the new technologies on a global level. In 2003–04, Indian biotechnology companies had combined revenues of more than US$700 million and they surpassed US$1 billion in 2005. According to the World Health Organization (WHO), India is the fourth largest producer of pharmaceuticals.[3] This phenomenon has generated investment in research and development in every field of science, including the life sciences and genomic research. 'Public–private partnership' is a new buzz-word in this field and several public genomic research institutes of repute have set up collaborative research programmes with leading national and multinational pharmaceutical and IT companies in throughput research areas.[4]

Increase in funds and infrastructure has pushed many research institutes to take up genomic research with vigor of late. In 2001, the Department of Biotechnology, Government of India, negotiated with many pharmaceutical and biotech companies for collaborative ventures in secondary genomic research, and the focus of these projects was in areas such as proteomics, functional genomics, data mining, curation, and annotation of genome data. The biotechnology boom coupled with India's rich biodiversity, a huge medical infrastructure, a large population size, and an expanding market has made India a hunting ground for genomic research. Issues such as the export of genetic materials,[5] unethical drug trials,[6] and the commodification of genetic materials called for regulation. Two central bodies, the Indian Council of Medical Research[7] and the Department of Biotechnology[8] formulated ethical guidelines and policies on the conduct of biomedical research on human subjects. With the rapid expansion of genomics-related research vis-à-vis the exposure of the diverse population groups that inhabit India, many issues surfaced on the handling, protection, and use of genetic information. This led the central government to set up a National Advisory Committee to look into the emerging issues. Apart from the issues related to its modalities, ownership, ethical, legal, and social implications, the NR will have to deal with the structural and functional aspects of its relationship with source institutes (those who will share genetic resources and data with the NR). At the same time, the NR will have to take into account its relation with collaborative partners, including private pharmaceutical and bioinformatics companies vis-à-vis the source institutes.

The planning of a NR is pregnant with bioethics- and policy-oriented issues in biological research involving human subjects, which are believed to transcend the interests of all stakeholders. With diverse interests at stake, the views of stakeholders concerning the NR differ as well. Thus, existing repositories have their independent opinion about the necessity and modalities of the NR.

Without going into those details, we will focus on the basics, i.e. the human genetic samples, people who contribute them, and the process of data gathering. It is evident that for the formation of a repository, genetic data are crucial. But what constitute genetic data and how they are generated or gathered in a particular societal context is of prime importance in the understanding of debates concerning bioethics in genomics. Many scientists argue[I] that the formation of an NR can stop the illegal supply of bio-samples to foreign labs, and avoid the setting up of many different biobanks of numerically small, vanishing tribes. Moreover, it could avoid repetition of the collection and study of samples from the same population groups, minimise costs and time spent on biobanking, and facilitate the central management of data.

With the advent of genomics and associated technology, a great deal of genetic samples and data including health data have been generated through population carrier-screening, human genome diversity studies, and other genomic-related research, along with hospital-based studies. A closer look at the process of data generation will shed light on the related ethical, legal, and social issues involved in genetic research crucial in the genomics debate in India.

The population carrier-screening programme: its role in genomics and biobanking

As mentioned above, a population carrier-screening programme for genetic disorders is one way of generating genetic samples and health data crucial for the establishment of a biobank. Of course, there is no direct link between the existing or proposed biobank or the NR and the population carrier-screening programmes, such as the carrier-screening programme for sickle cell anaemia at various sites in India. However, the programme provides an indirect database with both genetic- and health-related data, crucial for genomic studies. An indirect link also exists where selected subject participants from the community are sampled for genomic research, and are again taken as patients whose bio-samples are collected as follow-up data for genomic research.[II]

The purpose of discussing the carrier-screening programme in the context of biobanking in India is twofold. First, it provides a ground where a parallel comparison of data generation procedures vis-à-vis bioethical guidelines and protocols can be made between biobanking and carrier screening, as both deal with human subjects in biomedical research;[III] and second, it provides a basis for discussion on the ways in which genetic screening and biobanking are articulated in India by various stakeholders.[IV]

Population carrier-screening programmes for monogenetic disorders such as sickle cell anaemia, G-6-PD, and phenylketonurea (PKU) have been carried out for many years in India.[V] The main focus of such studies was to identify carriers in a given population. This would facilitate the generation of data on disease frequency for epidemiological studies and healthcare management. Initially, it was done basically as an academic pursuit. Then, with the findings of a high prevalence rate of carriers and disease for certain genetic disorders among a few defined population groups, genetic intervention studies were carried out with central and state government funding. Such studies at the same time provided pre-marriage genetic counselling[9] and treatments.[10]

Sickle cell anaemia is a hereditary abnormality of haemoglobin, which affects millions of people throughout the world. It poses a major genetic and public health problem in India, causing haemolytic anaemia, high morbidity, mortality, and reproductive wastage. It affects the general health, the psycho-social behaviour, and the economy of a large number of tribal as well as non-tribal peoples.[11] More than three hundred population groups of India belonging to different ethnic communities have been screened for the HbS (sickle cell disease) gene by a number of scholars.[12,13] State-wide distribution of the sickle cell gene clearly demonstrates that the concentration of the HbS gene is primarily localised in three pockets: (1) Tamilnadu and Andhra Pradesh; (2) Orissa and Madhya Pradesh, and (3) Gujarat and Maharashtra, besides the sporadic occurrence of the gene in other parts of the country.[14,15] The high prevalence rate of sickle cell disease among the Schedule Tribes (STs) is reported as a result of large-scale studies conducted by physical anthropologists/human geneticists from various universities and research institutes. Interestingly, although the rates are high among the STs, there are a few ST groups among whom the gene is absent. Another compilation of data from 179 different groups and subgroups of STs and Schedule Castes (SCs) in respect to the prevalence rate of the sickle cell trait shows that in 79 groups over 15 per cent of the population are carriers of HbS, and the prevalence rate is over 19 per cent among 35 tribal populations.[16]

Through our anthropological study on emerging issues in genetic sampling and data banking in India,[VI] we touch upon common bioethical principles that are debated in genomics and biobanking and their responses in India from a community perspective. We focused on four principles that we considered crucial: informed consent, confidentiality, benefit sharing, and public trust. We interviewed people from a cross-section of various stakeholders. In the following paragraphs we will discuss each of them on the basis of how they are understood in general, and on the basis of their relevance and importance in the context of India's biobanking initiative.

Social and ethical issues in biobanking and the Indian scenario

Biobanking has raised at least four major ethical issues pertaining to: (1) the tension that exists between the interests and the rights of individuals/groups in the progress of a scientific research project; (2) the provision of adequate information on the project to individuals before acquiring their consent to deposit their samples, as well as raising awareness of the way their samples and their associated data could be used in unforeseen research and studies; (3) the difficulty of reconciling the non-commercial use of human body parts with the growing role of commercial biobanks; and (4) the often conflicting activities of finding the optimal and transparent use of biobanks and defining the rights of priority of researchers and companies over samples and data.[17] The underlying issues that can be deduced from the above four points exhibit two crucial yet conflicting social reactions. On the one hand, there is very strong support for breakthroughs promising better medical diagnosis and treatments and, on the other, there are anxieties about the increased loss of privacy and the potential for genetic discrimination, as well as doubts about the capacity to regulate genetic science in the public interest.[18] For the latter, in the context of biobanking initiatives or the research use of human biological materials, two basic protections for individuals and their community are considered: the requirement of informed consent and the overseeing by institutional review boards (IRB) or ethical committees to ensure an acceptable balance between the risks and benefits of research.[19]

Informed consent and its limitations

The principle of informed consent, which is the paradigmatic expression of the patient's autonomy, has its origins in the Nuremberg Code of 1947, having thereafter been included in the Helsinki Declaration of 1964. At present, virtually all medical ethics and research codes hold that physicians and researchers must obtain the consent of patients and subjects prior to any substantial intervention.[20] Its importance has been reconfirmed and reinforced through genomic research and biobanking initiatives.

Informed consent has been a hallmark of the enrolment in research studies and in medical procedures over the last half a century. Accordingly, individuals are expected to have the opportunity to give informed consent on a voluntary basis and to make an educated decision about their treatment. In situations where the individual's capacity to decide is impeded – such as with young children, the mentally impaired, or those with severe physical trauma – a proxy decision-maker is often called upon. Public health poses additional challenges for the notion of informed consent, since here the intervention is applied to an entire group, not just to one individual. Who is to decide whether to allow a group of researchers to enrol the community in a research study, or to draw samples from the community's members? Should it be a selected member of the group or should all members decide for themselves on an individual basis? How does the process of informing or educating the entire group take place? Should the community be consulted on the design of the protocol? In fact, informed consent in the context of public health seems to be an evolving concept, with new ideas constantly being voiced in the literature.[21]

It is worth noting here that the requirement for consent to be fully 'informed' may be objectively impossible to achieve. Even when correct information is carefully presented in culturally appropriate ways, it cannot be guaranteed that it has been understood. The ethical obligations that are achievable include an accurate delivery of information, together with the disclosure of the relevant risks and benefits to the individuals and communities involved. The information should be conveyed in language understood by potential research subjects, and should take into account the particular cultural and religious needs and aspirations of communities.[22] Although in itself it does not protect an individual against the misuse of provided data, informed consent allows individuals to exercise their fundamental right to decide whether and how their body, and its parts and the associated data, will be used in research.[23]

In fact, in the Indian context, researchers frequently consider informed consent more as a tool than as a bioethical principle, used in the procedure of data collection to 'legally protect' the respective interests of researchers and subject participants. It is the impoverished subject participants, poor and illiterate, who often get persuaded or manipulated through the use of this tool by researchers or interest groups.[24] Different stakeholders have different opinions on and understandings of the significance of informed consent.

When asked about informed consent procedures, Dr MM, a professor of physical anthropology at an Indian university heading a project on human genome diversity study on four tribal communities from central India, said:[VII]

> It is our longstanding relationship with the people that is enabling us to gather samples, not any assurance or incentive. Informed consent as it is understood in the bio-ethical and legal field, is neither feasible to take, nor is it effective. It is peoples' faith and goodwill in us which is important …
> The procedure of individual consent is too risky and impractical in case of rural, tribal and illiterate societies. We receive their non-cooperation at the moment we ask for written consent. Usually it is a matter of taking their signature or finger print on a piece of paper, declaring that they have understood the intent, the objective, the risks and benefits involved in the research for which they provide genetic data. This happens because they have had bad experience of property dispute or money-lending associated with providing a signature and a finger print. Another issue is that it is morally problematic to ask for consent when the subject is not in a position to understand the inherent meaning of the issues and we ourselves are not sure about the future implication of the data (Case 1).

About the process of data generation and informed consent, Mr RK, a PhD research student jointly attached to a university department and a research institute that is working on the molecular aspects of genomic analysis, made the following comments:

> There is a set of guidelines that we were told to follow during the brainstorming session of the project's execution plan. But the real field situation is something different. You go to a tribal village where you do not get any kind of cooperation from the people who do not show any interest in your work … What we are given is research targets: to collect 50 or 100

samples in a week's time. It is unthinkable to establish the kind of rapport with the subjects that is seen when using typical anthropological fieldwork methods. Forget about educating and informing people about the risks and benefits involved, we use some trick to get their blood samples. Our aim is to obtain the blood sample. Of course we take maximum care not to misuse it. We normally take a medical team with us and declare in advance that a medical team will come to the camp near your village and will check your general health status, test your blood and distribute free medicine. According to the guidelines, we are required to take samples from unrelated individuals, but that is not possible in a camp situation. What we do is take blood samples from all individuals and collect the genealogy and other family details. And then we select the samples that we need for our purpose, and the rest we throw away. We know it is not ethically correct, but we do it, as it is very difficult to get people's support and their samples in less time (Case 2).

Dr PM, a teacher of biological anthropology and one of the leading figures in genomic research and also a member of numerous ethics committees in India, was more categorical about informed consent and other bioethical issues. He opined:

We give maximum importance to informed consent in the data collection procedure. Written consent is the most preferred way; however, with the given population context, we do rely on oral consent with proper audio and video recordings. Now we are even thinking of using this method for all the subject participants. We even think of having the entire community consultation recorded . . . We take maximum care in maintaining the confidentiality of the subject participants, even the name of the community we don't reveal to others and we do not even publish it. All the samples are coded and strict confidentiality is maintained in this regard . . . As a member of a national committee, I pleaded for the benefit to the participant subjects and communities. I argued for five percent of the research profit to go to the community and participants. But the detailed modalities are not discussed . . . I will be glad if you can show me an instance of benefit being shared with people from anywhere in the world. We in India have not reached that stage where benefit is seen in tangible terms (Case 3).

The interviews above provide an idea about the way in which informed consent is practiced in India. It gives a glimpse of the gap between the official position based on ideals of bioethical protocols and the actual practice followed in the field. Dr MM (Case 1) represents the official position, whereas Mr RK (Case 2) and Dr PM (Case 3) represent the field practice or situational dimension. These stated conflicting ways reaffirm the urgency of the question of whether the principle of informed consent is 'doable' and 'desirable' in the context of developing societies, especially among tribal and small-scale communities. It also opens a debate on the problems of standardisation of guidelines in data generation procedures and the effectiveness of institutional review boards in the streamlining and the assessment of ethical clearances.

Group consent: its importance and vulnerability

Informed consent has raised much controversy regarding the ethics of human population genetics. Although some have questioned the application of even individual informed consent in some stages of research, the more interesting question, and the source of a burgeoning academic literature, is whether some other form of consent should be required from the group as a whole.[25] The requirement for group consent has been recognised by the Tri-Council of Canada,[26] Taiwan Biobank,[27] and by the Human Genome Diversity Project,[28] which supports the principle that population consent, as well as individual consent, should be sought for genetic research.[29]

Those in favour of community consent use it either in place of individual consent or as supplementary to it. Group consent refers to the ways of protecting communities through the consultation between researchers and communities or its leaders, before the drafting of a project, during its implementation, and during the final version of the draft. The question is whether the community allows the researcher, having consulted with the community, to proceed with a proposal, even if the researcher ignores the community's recommendations. In principle, the community has the power to veto the proposal. But, in practice, community consent can be restrictive of individual freedom. For this reason it is in great need of justification.[30]

In an Indian context, the concept of community consent is borrowed from traditional anthropological fieldwork techniques to achieve quick penetration of the community by obtaining approval for the recruitment of subject participants without antagonising individual participants. In practice, it neutralises the importance of individual informed consent procedures and hastens the data gathering process. As one researcher opined:

> If we plan to take some samples from a community, we first approach the village or community headman, the local teacher, the local representative or key person who can convince their own people, act as a buffer if there is any resentment, give some kind of protection to use and facilitate the smooth conduct of research. Once they give the permission, getting individual support becomes very easy. It is crucial in community research.

About community consent, Dr MM said:

> In the generation of genetic data we just have informal consultation with the village/community leaders such as the village headman, the local school teacher or the local level Panchayat representatives. Though we give equal

importance to individual consent, at times, with the approval and consent of community leaders, you can ignore or bypass the individuals.

Concepts such as 'community engagement' and 'community consultation' are ethical watchwords used in population-based studies of human genetic variation, where the theoretical aim is to allow human populations, i.e. the subjects of research, to have a degree of meaningful control over the initiation and conduct of a study. The theoretical aim of community engagement or consultation is to enhance the population's control over the ways in which its members are studied.[31] However, in practice the notions of community engagement and community consultation become a strategic tool for the recruitment of individuals as subjects of genetic research.

Confidentiality

The concept of confidentiality is based on the notion of privacy, which in certain contexts is a fundamental human right deserving of respect. The protection of privacy in the context of genetic research can be particularly challenging, for reasons related to the uniqueness of the individual's genetic code. The effect of the revelation of an individual's genetic information is not limited to the present condition of the individuals concerned in the study, but may also affect their family, their own future health and the health of their progeny. For these reasons, the retrieval, banking and use of genetic information must be done in a way that assures a satisfactory degree of confidentiality to all affected individuals.[32] Indeed, confidentiality in genetic biobanking covers aspects of, for instance, individual privacy, community privacy, data protection, access to and the sharing of data.

One way of ensuring confidentiality for individuals would be to separate identifying information from the national repository. As this could limit the scope of research, alternatively, the confidentiality of identifying information in the central repository could be safeguarded through coding and anonymisation.[33] However, it would be unrealistic and unwarranted to request that genetic information be kept in complete anonymity. It is not technically feasible in the first place, but it also deprives research participants of the opportunity to withdraw from a research project and it keeps them from receiving individualised research results when justified.[34] In the Indian context, very little can be said about how a national repository will affect the privacy of individuals. However, there are many instances in genetic research among communities in which individuals have been put at risk of social stigma and discrimination. In the absence of universal healthcare, life insurance, and full employment in the organised sector, the threat from insurance and employment in India is not as high as it is from social stigma and discrimination.

It is more difficult to maintain confidentiality at the community level, a situation best characterised as a double-edged sword. On the one hand, communities that are found to have a gene that predisposes them to a common late-onset disease, for medical reasons, may need to be identified. Their entering into a healthcare intervention programme may well be an impediment to the protection of their identity. Also, from the researcher's point of view, a particular ethnic tag facilitates the interpretation of the biological significance to be used in future studies of genetic association, population history, and relatedness.[35] On the other hand, any link made between genetic inferiority and individual identity or ethnic tagging will lead to the stigmatisation and discrimination of individuals and groups in areas such as insurance, employment, education, and social interaction.

In India, damage has already been done through epidemiological research. A few defined population categories, especially tribal and small-scale caste communities, are linked with certain genetic disease profiles, sickle cell being one such example. At times, as a result of genetic profiling, the stigmatisation of individuals and ethnic groups leads to social and psychological apathy. As can be seen among the Agaria caste group in the Sundargarh district of the state Orissa, which has a 20.5 per cent prevalence rate for the sickle cell trait, the entire community is known to be prone to this disease. Neighbouring caste communities decry the unethical and immoral marriage practices of the Agarias. As the Agaria constitute a small community and traditionally marry among close relatives, their susceptibility to sickle cell disease has made them vulnerable in the eyes of their neighbours. As one young Agaria man said:

> It is not that we Agarias all carry the sickle cell trait or that we all suffer from the disease, but other caste people think that every Agaria is a sickle cell patient and it is because we practice consanguineous marriage. It is pathetic … People do not understand that only a few have this trait and very few among them are sufferers. This disease is found among many other communities.

Who can have access to the genetic information and materials in biobanks is a contentious issue in genomics and genetic research, especially as the legal and cultural authorisation varies across societies, cultures, and nations. Some arguments favour access to extend to family or 'close blood relatives', whereas current guidelines uniformly prohibit the disclosure of genetic information to third parties without the express consent of the participant.[36] Our study in India, in the context of genetic intervention studies on sickle cell anaemia, shows how access to genetic information does not automatically honour individual and group privacy, and can lead to disclosures that stigmatise participants and their families. For instance, as part of its genetic intervention programme, the Raipur branch of the Indian Red Cross Society carried out a population carrier-screening programme in a village *Panchayat* (local government) in Raipur district [name withheld]. In the village, a two-day screening camp was conducted with the help of village committee members. A few days later a computerised list of carriers and sufferers was given to the village committee members for distribution, not to the concerned individuals or their family, grossly violating the principle of confidentiality and privacy. Even though a copy of the list is held with one of the authors, the Society official blatantly denied its existence in a subsequent interview.

Benefit sharing

Although the commodification and the commercialisation of the human body and its elements are not considered ethically sound, these have not been brought to a halt. As if this ethical dilemma can be compensated for, an emerging approach hopes to provide some return of benefits to the community.[37] UNESCO's 'Universal Declaration of Bioethics and Human Rights' 2005, the Human Genome Organisation's *Statement on Benefit Sharing 2000*, and other guidelines mention the principles of benefit sharing in general, and in particular the equitable distribution of benefits from research. UNESCO's 'International Declaration on Human Genetic Data' is one of the most emphatic assertions of the principle and states that 'benefits … from the use of human genetic data … should be shared with the society as a whole and the international community'. However, the principle is disconcertingly amorphous, particularly in relation to the operation of intellectual property protection and licensing.[38,39]

Much current genomic research is of a tentative and explorative nature, and (beneficial) research results can often not be predicted. The key question is, then, what can be considered as, and whether there is a reasonable chance of, future benefit.[40] At least at present, research in the field of (pharmaco-) genomics generally is non-therapeutic in nature and aims at furthering societal benefits through increased scientific and medical knowledge. Only very occasionally can it offer benefits in the form of therapy or prevention. The bigger challenge will be to predict and assess the particular modes of benefit so that prospective subject individuals, patient groups, and participant communities can be informed about their share in the expected benefit. Participant individuals and groups can have a legitimate claim only when their participation is acknowledged or properly documented. With the kind of mechanism at work in India at ground level, where informed consent is not maintained properly and the identity of individuals is not automatically recorded or maintained, it is highly unlikely that justice can be ensured as far as benefit sharing is concerned. Though existing protocols (ICMR 2000 and DBT 2002)[7,8] talk about benefit sharing, the mode and mechanism of sharing between unequal players (powerful authority and less powerful subjects) is not fully discussed. DBT (2002) guidelines say 'it is obligatory for national/international profit making entities to dedicate a percentage (e.g. 1%–3%) of their annual net profit arising out of the knowledge derived by use of the human genetic material, for the benefits of the community.' A prominent scientist and active member in many committees says:

> Though the issue of benefit [mostly financial gain out of research results] is in a conceptual stage, we need to talk and devise methods so that justice

is guaranteed, but for that we need to have instances of benefit sharing in other countries. We in India have not reached that stage in genomic research in which benefits are tangible, so that a party can make a claim in the first place.

At issue here is the question of what a benefit is and how participant individuals and communities view it in their own socio-cultural context. Another area of contention is how to determine who is a beneficiary. The issue of the ownership of the data (i.e. bio-samples) and the legal authority over the samples is problematic. Whether it is an individual, a family, a patient group, or a community that holds the authority over the sample is a very tricky and complex issue. Paradoxically, the idea of sharing of benefit has created a poignant sense of moral conflict. Benefit sharing seems to sanction the commodification of human body parts, devaluing the worth of human life and failing to recognise human dignity.[41]

Public trust and biobanking

It is generally believed that a human biobank is an essential tool,[42] has great potential for biomedical research,[43] and may lead to therapeutic discoveries that can augur healthcare improvement. The initiative of setting up a biobank rests upon the trust of patients and healthy persons donating blood and samples,[44] and also on the trust of the broad masses of the public. It is the exclusion of patients from information about the purpose of human tissue sampling and the destination of the samples that has caused much of the public debate on biobanking. It may be that public trust can only be restored through a policy of transparency as to how the multiple interests of current and future patients and subject participants are recognised and handled. Institutional review boards (IRB) and ethical review boards are expected to play a key role in maintaining public trust.[45]

With the rapid development in biotechnology and bioinformatics, biobanking initiatives are expanding globally with an added focus on the commercial exploitation of research results. Public and private partnership in research and development has become an established norm in this new field due to an increasing demand for investment in technology and manpower, and the widened scope for applications leading to commercial benefit. Just as commercialisation has challenged public trust, so too public trust is an essential condition for biobanking. The Human Genetic Commission of the UK, for example, is careful to promote realistic expectations of the pace of scientific and medical research and the role of partnership between public and commercial research[46] citing the risk in declining levels of public trust. *Nature* observes:

Researchers who work for profit-making ventures should be particularly careful, as their motives are more open to suspicion. In a poll taken in March 2002 for the British Royal Society, more than half of the respondents thought that science funding is becoming too commercialized, and said they wanted more influence over the research agenda. Scientists will keep the public's trust as long as they consider citizenship as well as scholarship.[47]

In the Indian context, public trust in the field of biobanking will be crucial as there already exists distrust among the public about the new technologies and their applications. Now, people are becoming increasingly aware of the technological complexities, making efforts to stimulate public debate on the subject imperative.

Discussion

In India there are three areas of concern crucial in shaping the public perception of the role of genomics and bioinformatics in improving health conditions: First, the country's health needs and current priorities in healthcare research, in particular the emphasis on genomics; second, the rationale and purpose of research and the maintenance of bioethical protocols; and finally, health inequity and the scope of public–private partnerships in research and the resultant commercialisation of research results.

The public needs to be informed about the prioritisation of public funding for healthcare research and development, on which India spends less than 0.5 per cent of its total budget,[48] compared to a defence budget of 2.68 per cent of GDP.[49] An estimated four million people are living with AIDS, 420,000 die annually from tuberculosis, and diseases such as cancer, diabetes, and heart diseases are becoming considerable healthcare concerns. The Indian biotechnology industry is ranked third in the world in terms of stem cell research, primarily because both the government and private industry have invested heavily in research institutes studying stem cells.[50] Here, the public needs to be informed and educated about the prioritisation and balancing act that India needs to strike in furtherance of research and in helping to meet the health needs of economically weaker sections of society.

In a biobank initiative, the participation of patients and healthy individuals in providing bio-samples and health data is crucial. At the same time, the biobank authorities demand compliance with higher scientific standards of procedure in data collection, compilation, storage, and manipulation. A balance needs to be struck between informing and educating the subject individuals and communities about the objectives of the project, consent procedures, confidentiality, and coding mechanism and terms of benefit sharing, on the one hand, and maintaining the independence of researchers and scientific values, on the other. A main concern here is minimising the gap between in the bioethical guidelines and protocols and the practice at ground level. For example, the IRB could perform mid-term assessment of the credibility of an ongoing research and its adherence to bioethical principles.

In India, the biotechnology industry is capital intensive and the funds required for biomedical research can often be mobilised only with the help of the private sector;[51] and there has been a significant amount of collaborative research between leading genomic research institutes, universities, bioinformatics and pharmaceutical companies.[52,53,54] However, the public has very little access to information on the exact terms and references of these collaborations, while a tendency of corporate secrecy and bureaucratic fatigue render expectations of immanent transparency misplaced.

Acknowledgements

The research for this article was conducted within the framework of the research project *Socio-genetic Marginalization in Asia Programme (SMAP)* financed by the Netherlands Organisation for Scientific Research (NWO) at the IIAS, Leiden, the Netherlands. The authors would like to thank all the respondents in the field who participated in the research. Also thanks are due to Dr Jyotsna A. Gupta for her comments and inputs on an earlier draft of this paper. Address for correspondence: Dr Prasanna Kumar Patra, Research Fellow (SMAP), International Institute for Asian Studies, Leiden University, PO Box 9515, 2300 RA Leiden, The Netherlands. Tel: +31-71-527-2419, Fax: +31-71-527-4162, e-mail: pkpatra@rediffmail.com.

Notes

1 Jayaraman, K.S. (2004) India aims to become the main bioinformatics hub, *Nature Biotechnology*, 22(8): 933.
2 Acharya, T., Kumar N.K. *et al.* (2004) 'Harnessing genomics to improve health in India' – an executive course to support genomics policy, *Health Research Policy and Systems*, 2:1.
3 Jayaram, K.S. (2005) Biotech boom, *Nature Biotechnology*, 23(9): 1183–4.
4 Bioinformatics – India discovers goldmine (2002) http:www.indianpurchase.com/magonline/purchase/200208/article1.htm (downloaded on 29 December 2005).
5 Mudur, G. (1996) India concerned at export of genetic material, *British Medical Journal*, 312: 464.
6 Srinivasan, S. (2004) Indian guinea pigs for sale: outsourcing clinical trials. India Resource Center. http://www.indiaresource.org/issues/globalization/2004/indianguineeapigs.html
7 ICMR (Indian Council of Medical Research) (2000) Ethical guidelines for biomedical research on human subjects. *ICMR Bulletin*. 30(10): 107–16. Also http://www.icmr.nic.in/vsicmr/ethical/pdf
8 DBT (The Department of Biotechnology, Government of India) (2002) Ethical policies on human genome, genetic research and services (http://dbtindia.nic.in/policy/polimain.html).
9 DBT. See website: http://dbtindia.nic.in/publication/publicmain.html
10 Kate, S.L. (2000) Health problems of tribal population groups from the state of Maharashtra. http://sickle.bwh.harvard.edu/india_scd.html
11 Balgir, R.S. (2001) Genetic epidemiology of sickle cell anemia in India, *Current Medical Trends*, 5:872–80.
12 Basu, S. (1994) *Genetic Disorders and Health Care*. Delhi: Shree Kala Prakashan.
13 Balgir, R.S. (2000) Human genetics, health and tribal development, In: Sharma, P.

Dash (Ed.) *Environment, Health & Development,* Ranchi, Ray Institute of Anthropological Studies.

14 Mukharjee, B.N. and Das, M.K. (1990) Spatial distribution of two predominant abnormal hemoglobin HbE and HbS in India sub-continent, *Journal of the Indian Anthropological Society*, 25: 39–59.

15 Das, P.K. (2001) Bio-Anthropological Perspective of Sickle Cell Disease in India. Anthropology and Archaeology Sectional Presidential Address, 89th ISCA Session 2001–02.

16 Basu, S. (1994) *Ibid.*

17 Cambon-Thomsen, A. (2004) The social and ethical issues of post-genomic human biobanks, *Nature*, 5: 866–73.

18 Knopper, B.M. (2004) Consent and confidentiality in population genetics and epidemiological research: ethical aspects of biobanking in an international context – is ethics promoting or preventing sharing? PowerPoint presentation, GenomEUtwin Ethics Educational Meeting, Oslo, 21August.

19 Godard, B., Schmidtke, J., Cassiman J.J., *et al.* (2003) Data storage and DNA banking for biomedical research: informed consent, confidentiality, quality issues, ownership, return of benefits. A professional perspective, *European Journal of Human Genetics*, 11(S2): S88–S122.

20 Andorno, R. (2003) Seeking common ground on genetic issues: the UNESCO declaration on the human genome. In: Sandor J. (Ed.) *Society and genetic information: Codes and laws in the genetic era*, Budapest: Central European University Press.

21 Key Points on Ethical, legal and social issues in public health genetics (PHELSI). (See: http://www.sph.umich.edu/genomics/media/subpage_autogen/PHELSI.pdf).

22 Leng C.H., El-Hamamsy L. *et al.* (1995) Bioethics and human population genetic research. United Nations Educational, Scientific and Cultural Organization and International Bioethics Committee. CIP/BIO/CONF.002/5.

23 Sade, R.M. (2002) Research on stored biological samples is still research, *Archives of Internal Medicine*, 162: 1439–40.

24 Patra, P.K. and Sleeboom-Faulkner M. (2006) Genetic bio-banking in India: ways and means in data generation – a community-based perspective. Paper presented at 2006 ELSI Symposium on Reexamining the ELSI Implication of Biobanking – A Cross Culture Perception, held at Taipei 17–18 September 2006.

25 Greely, H.T. (2001) Informed consent and other ethical issues in human population genetics, *Annual Review of Genetics,* 35: 785–800.

26 Tri-Council of Canada: Medical Research Council, Natural Sciences and Engineering Research Council: Social Sciences and Humanities Research Council, (1997) *Code of Conduct for Research Involving Humans*, Ottowa, The Ministry of Supply and Services Canada.

27 Tsai, D. (2006) Mediating media effects within public participatory opinion in structuring Taiwan Biobank. Paper presented at 2006 ELSI Symposium on Reexamining the ELSI Implication of Biobanking – A Cross Culture Perception, held at Taipei 17–18 September 2006.

28 North American Regional Committee: Human Genome Diversity Project. (1997) *Houston Law Review*, 33: 1431–73.

29 Godard, B., Schmidtke, J., Cassiman, J.J., *et al.* (2003) *Ibid.*

30 Wilkinson, T.M. (2004) Individualism and ethics of research on humans, *HEC Forum*, 16(1): 6–26.

31 Juengst, Eric T. (2003) Community engagement in genetic research: the 'slow code' of research ethics? In B.M. Knoppers (Ed.). *Populations and Genetics – Legal and Socio-Ethical Perspectives*, Leiden: Martinus Nijhoff.

32 Deschenes, M., Cardinal G., Knopper B.M., *et al.* (2001) Human genetic research, DNA banking and consent: a question of 'form'? *Clinical Genetics*, 59: 221–239.

33 Leng, C.H., *et al.* (1995) *Ibid.*
34 Joly, Y. and Knopper, B.M. (2006) Pharmacogenomics data sample collection and storage: ethical issues and policy approaches, *Future Medicine*, 7(2): 219–26.
35 Indian Genome Variation Consortium (2005) The Indian Genome Variation database (IGVdb): a project overview, *Human Genetics*, 118: 1–11.
36 Deschenes, M., *et al.* (2001) *Ibid.*
37 Knopper, B.M. (1999) Status, sale and patenting of human genetic material: an international survey, *Nature Genetics*, 22: 23–6.
38 Chadwick, R. and Berg, K. (2001) Solidarity and equity: new ethical frameworks for genetic databases, *Nature Review Genetics*, 2: 318.
39 Chalmer, D. (2006) *Ibid.*
40 Hoedemaeker R., Gordijn, B., Hekster, Y., *et al.* (2006) The complexities of ethical evaluation of genetics research, *HEC Research*, 18(1): 18–36.
41 A benefit is a good that contributes to the well-being of an individual and/or a given community (region, tribe, disease groups, etc.). Benefits transcend the avoidance of harm (non-malfeasance) in so far as they promote the welfare of an individual and/ or a community. Thus, a benefit depends on various needs, values, priorities, and cultural expectations. HUGO Ethics Committee (2002) Opinion: Hugo Ethics Committee statement on benefit sharing, *Clinical Genetics*, 58: 364–6.
42 Chalmer, D. (2006) *Ibid.*
43 Kaiser, J. (2002) The population database boom, from Iceland to the US, *Science*, 298: 1158–61. http://www.genomics.ee/files/artiklid/science_08_11_02.pfd
44 Hansson, M.G. (2005) Building on relationships of trust in biobank research, *Journal of Medical Ethics*, 31: 415–18.
45 Hansson, M.G. (2005) *Ibid.*
46 Chalmer, D. (2006) *Ibid.*
47 Anonymous, (2002) Trust and how to sustain it, *Nature*, 420(19/26): 6917, http://www.nature.com/nature/journal/v420/n6917/full/420719a.html
48 Acharya, T., Kumar N.K., Muthuswamy, V., *et al.* (2004) Harnessing genomics to improve health in India – an executive course to support genomics policy, *Health Research Policy and System*, 2: 1.
49 India's defense budget 2000–2001. http://www.indianembassy.org/policy/defense/india_defense_budget_2000.htm
50 Acharya, T., *et al.* (2004) *Ibid.*
51 Acharya, T., *et al.* (2004) *Ibid.*
52 Narendranath, K.G. (2001). Medical biotechnology – A DBT vision for India to attain new heights, *Pharmabiz.com*, 11 October 2001. http://www.pharmabiz.com/article/detnews.asp
53 Ghungurde, A. (2000) Human genome project whets pharma firms R&D appetite, *Financial Express,* 8 March.
54 Indian Genome Variation Consortium (2005) *Ibid.*

Notes

I During Patra's fieldwork, this came out in his discussions with many scientists and researchers who are actively involved in genomic research and biobanking and who are aware of the issues related to the establishment of the national repository.

II The booklet on sickle cell anaemia published (in English and Hindi) and distributed by the India Red Cross Society, Raipur Branch, says that the Society wishes to carry out a molecular study of sickle cell anaemia based on its research and intervention programmes in the state in collaboration with throughput research institutes such as the Center for Chemical and Molecular Biology (CCMB), Hyderabad. During fieldwork, it was revealed that the District Medical College and Hospital are establishing a laboratory for such study and initiating a genetic counselling programme in collaboration with the Society.

III Indian Council of Medical Research (ICMR)'s guideline *Ethical Guidelines for Biomedical Research on Human Subjects (2000)* and Department of Biotechnology (DBT)'s guideline *Ethical policies on Human Genome, Genetic Research and Services (2001)* talk about ethical and legal principles that need to be followed in all kinds of genetic and biomedical research involving human subjects.

IV Here, stakeholders include, among others, individual subject participants and their community who provide bio-samples and health data as resource persons, researchers and scientists along with their institutions who are involved in such data generations, policy-makers and authorities as owners, and decision-makers and companies and collaborators who have commercial and other stakes in research.

V Population-based genetic carrier-screening programmes, especially for mono-genetic disorders such as sickle cell anaemia have been carried out among various communities, especially tribal communities, for many years for academic purposes. For details, see the compilations of Bhasin M.K., Walter, H. and Danker-Hope, H. (1994). *People of India: an investigation of biological variability in ecological, ethno-economic and linguistic group.* Delhi: Kamal-Raj Enterprises; and Basu, S. (1994). *Genetic Disorders and Health Care.* Delhi: Shree Kala Prakashan.

VI Through the project *Emerging Issues in Genetic Sampling and Data Banking in India – a community based perspective* (under the broad framework of *Socio-genetic Marginalization in Asia Programme*) we carried out research on ELSI issues in genetic sampling and biobanking in India among several tribal communities from Orissa, Chhatishgarh, Maharastra, and Gujarat states and at various universities, research institutes and hospitals across the country, and interviewed people from a cross-section of stakeholders using anthropological qualitative research techniques (for details visit www.iias.nl/smap).

VII The interviews were conducted in Hindi, English, and other regional languages and the translation was done by PKP.

9 Within borders

Risks and the development of biobanking in China

Wen-Ching Sung

Introduction

Biobanking is not a new practice, but since the initiation of the Human Genome Project, there has been a global boom in biobanks (Cambon-Thomsen 2004). Biobanks refer to collections of biological samples and attached databases, and come in many different forms. The data may be related to a given individual (clinical data), to a family (genealogy information or ethnic origin) or to a group (population living in the same geographical location or using the same language). A survey of biobanks in 147 institutes in six countries of the European Union (EU) shows that biobanking is increasing in all countries because activities related to genetics are proliferating (Hirtzlin *et al*. 2003). Biobanks may be used for research and technological development, medical diagnosis, or therapeutic activities. Although large population-based collections only represent one type of biobank, they have drawn attention from the public and the media. More and more large-scale biobanks have been established, and they become the icons of biobanks (Hirtzlin *et al*. 2003).

Changes in the practices and the significance of biobanks are directly related to the concept of 'genetic resources'. In both developing and developed countries, biobanking has become a way to appropriate their genetic resources for developing biotechnology and promoting public health (Cambon-Thomsen 2004). In other words, biobanking is much more than a means to retrieve scientific data for academic research; it has been advocated as an effort towards preserving national resources.

While countries establish biobanks as a way to harness their resources, the social status of biobanks is changing. Biobanks used to be solely clinical or academic research tools ignored by the public, yet now they have become a subject for debates within and between nations. In Europe and North America, people worry that the building of biobanks may increase the intrusion of the government into the private domain. Interestingly, if we make a cross-cultural comparison of the ethical issues involved in biobanking, then we can see a significant distinction between China and some Western countries. In China, the ethical concerns about privacy in biobanking are not the focus of public attention. By contrast, in public discourse, the danger of biobanking is often perceived as the danger of 'biopiracy', i.e. unjustified appropriation of the nation's genetic resources by foreign scholars or pharmaceutical companies. Therefore, the related legislation in China stresses the need to keep its human genetic resources within its borders.

This chapter contextualises the development of large-scale population-based biobanks in China. The author argues that the distinct ethical discourses on biobanking in China and in the West show the plasticity and constructiveness of our concepts about 'risks'. As sociologist Ulrich Beck indicates, risk is not a fixed, straightforward attribute of an action that can be directly perceived or estimated. Rather, risk is a product of deliberation: it has to be established through arguments (Beck 1992). From this perspective, the fact that the Chinese do not discuss the risk of biobanking in threatening individual privacy as much as Westerners does not mean that the Chinese fail to perceive such a risk. Neither does it imply that the consciousness of risk is absent in China or that the Chinese lack an awareness of individuality. Rather, it shows China's and the West's different ways of constructing the risk discourses on biobanking. More specifically, the Chinese government and the media perceive science differently, and tend to prioritise different kinds of risks. Furthermore, issues concerning biobanks in China reflect the lack of regulation for the global protocol of data collection.

This chapter will also examine how the concept of 'resources', derived from mines, animals and plants, affects how the Chinese government perceives genetic resources. For scientists, genetic resources embody valuable *information*. Yet the legislation of the Chinese government issued in 1997 focused on the *materiality* of the bio-resources. This contrast reveals both the bounded and non-bounded characteristics of the genetic resources.

Large-scale biobanking

The rise of large-scale population-based biobanks is one of the features of life science in the post-genomics era. The implementation of the Human Genome Project (HGP) heralded the beginning of a new discipline, genomics. The major aim of the HGP was to determine the exact order of the subunits of human genetic materials. Yet as anthropologist Paul Rabinow points out, it is recognised that the human genome can be changed. From the start of the Human Genome Project, 'representing and intervening, knowledge and power, understanding and reform, are built in, as simultaneous goals and means' (Rabinow 1999a).

Since the planning phase of the HGP, some scientists claimed that the human genome is the 'code of life,' and by 'decoding' the 'code of life' the project would help to discover the relation between genes and diseases. The meaning of biological samples therefore changes dramatically in the post-genomics era. Now many people hesitate to give blood or other samples because it contains all the genetic information that can be used as a basis for personal identification, for prediction of the possibility of having specific diseases, and for confirming biological relations among people. It is the promise of such powerful information that turns biological samples into valuable resources.

This scientific rhetoric of the 'code of life' made scholars alert to the underlying genetic determinism. Biologist Richard Lewontin wrote *Biology as Ideology* to criticise genetic reductionism. In the book, Lewontin urges us to consider the impacts of environment upon the development of organisms (Lewontin 1992). Science historian Evelyn Fox Keller points out that the genetic determinism delivered by molecular biology will not be self-contained. Instead, there will be an ideological expansion of molecular biology into both popular culture and medicine, leading to a tendency for the geneticisation of diseases and even behaviour (Keller 1992).

Interestingly, around the time of the completion of the HGP, the voice of scientists who emphasised the interplay between genes and environment became increasingly loud. More and more studies are devoted to examining the effect of the interaction between environment and genes on health. Prospective cohort studies such as the proposed biobank in the UK arise as a means to investigate the genetic and environmental factors that cause common diseases. A cohort study is a type of longitudinal study used in medicine and social science, in which a cohort refers to a group of people with common experience or characteristics. For example, a group of people who were born in a particular year may form a birth cohort. One specific advantage of cohort studies is to allow researchers to establish temporal relationships between cultural or biological exposure and outcome. In order to establish such temporal relationships, it is crucial for researchers to stay in touch with all the members of the group. Naturally, a cohort study is both time and money consuming. The UK biobank then is a landmark of such a study. It proposes to recruit up to 500,000 volunteers aged 40–69 for the study. The biobank stores DNA samples from which scientists hope to retrieve detailed information about lifestyle and medical history.

Biobanks in China

In the early twenty-first century, China launched two large-scale population-based biobanks: the Kadoorie Study of Chronic Disease in China (KSCDC) and the Guangzhou Biobank Cohort Study (GBCS). Both programs are prospective cohort studies for understating common chronic diseases, and are partly modelled on the UK biobank. The initiation of these programs reflects the concerns of the epidemiological transition in China. Due to urbanisation, economic development, and demographic change, China has undergone a rapid transition in the main disease patterns among its population. There has been a significant decrease in maternal and child mortality and in adult mortality from the main infectious/parasitic diseases. Meanwhile, death caused by non-communicable chronic diseases continuously increases. Such a change in the disease profile has caught the attention of Chinese scholars and the government. The biobanks take into consideration this new epidemiological situation.

Below, the author will analyse how the Chinese respond to these biobanks differently from the way Europeans react to their biobanks. The differences, it will be argued, reflect the risk issues surrounding biobanking. Let me introduce some features of these two biobanks first. The KSCDC aims to enrol 500,000 adults, the same size of the proposed biobank in the UK, between 2004 and 2008. The GBCS plans to recruit 50,000 residents in Guangzhou between 2003 and 2007. As of 2006, the GBCS had already enrolled 10,000 subjects, and published several scientific articles based on the analysis of this database (Jiang *et al.* 2006; Lao *et al.* 2006). Although KSCDC and GBCS are similar to the biobank project in the UK, unlike their counterpart in the UK and other European countries, these projects received little attention from Chinese and foreign mass media. Most interestingly, no protest or debates about these Chinese programs have ever been reported. The implementation of these projects has been smooth; and researchers plan to take only four to five years to complete the sample collection.

KSCDC is one of the largest blood-based prospective studies ever conducted in the world. The KSCDC is an open-ended prospective study with broad research aims. One of the research goals is to monitor the growth of the tobacco epidemic over the next decades in China, given the recent large increase in cigarette consumption among the population (Chen *et al.* 2005). It recruits adults aged 35–74 years old from the general population in ten geographically defined regions across China, including five urban areas (Qingdao, Harbin, Liuzhou, Suzhou, and Haikou) and five rural areas (Sichuan, Zhejiang, Hunan, Gansu, and Henan). For the researchers, this geographic diversity symbolises differences in the ways people live rather than genetic differences. The KSCDC is not the first large prospective cohort study of epidemiology in China. What is new about this project is that it collects both plasma and buffy coat samples from its subjects (Chen *et al.* 2005). By storing these biological samples, the study aims to assess the relevance of many genetic and other factors in various common chronic diseases. The major funding for the project came from the Kadoories Charitable Foundation (KCF) in Hong Kong. The UK Medical Research Council, the British Heart Foundation, and Cancer Research UK also provide core funding to the Clinical Trial Service Unit (CTSU) at Oxford University.[1] Even though its sample collection is of such a large scale, the KSCDC program has been extremely low-profile in China and in the international mass media. What is better known is the GBCS program.

In Chinese news reports, the GBCS is called the first biobank in China. Strictly speaking, the phrase 'first biobank' does not correspond to reality because other kinds of biobanks such as blood banks and banks for cell lines have been set up earlier than the GBCS. It is indisputable, however, that GBCS is the first large-scale population-based biobank in China. The GBCS was implemented by an international network among public health scholars: Tai-Hing Lam (Hong Kong), Chaoqiang Jiang (Guangzhou), and Kar Keung Cheng (Birmingham). The project was jointly funded by the University of Hong Kong, the University of Birmingham, Guangzhou No. 12 Hospital

(Guangzhou Occupational Disease Prevention and Treatment Center), the Guangzhou Public Health Bureau, and the Guangzhou Science and Technology Bureau (Jiang *et al.* 2006).

The GBCS recruits senior citizens aged at least 50 years old in Guangzhou, and plans to follow these subjects for 5 to 10 years. For the researchers, this cohort has been selected because they are more likely to develop chronic diseases within five to ten years. Given their relatively stable lifestyle, it is easier for researchers to maintain contact with these people (Jiang *et al.* 2006).

Guangzhou, the provincial capital of Guangdong Province in southern China, is one of the most economically developed regions of China, and thus is considered an ideal site for studying urbanisation in relation to chronic diseases. The total population of Guangzhou is around 10 million, which is composed of 6.4 million permanent residents with locally registered households and 3.4 million immigrants from other parts of country. The members of the GBCS's targeted group had experienced two major social transformations: the communist reform and the economic reform since the late 1970s. For researchers, the old urban residents in Guangzhou share a uniform 'Cantonese heritage with a common dialect, culture, and cuisine' (Chaoqiang Jiang *et al.* 2006:844). The objective for the researchers is to examine the roles of genes, lifestyle, occupational and environmental factors, and life course in causing the common chronic diseases.

The data collection practice of the GBSC opens a window for us to analyse the concepts of 'risks' and 'resources' involved in biobanking. To protect the participants, the study acquired ethics approval from the Guangzhou Medical Ethics Committee of the Chinese Medical Association, Guangzhou, China. Like most biobanks, the GBCS utilises a hospital as the central platform for collecting samples. Subjects come to the Guangzhou Number 12 Hospital, and are interviewed and examined by trained nurses and technicians. After taking fasting blood samples, the subjects pass through pre-set stations for interviews and a medical examination, including measurement of blood pressure, anthropometric indices such as weight and height, 12-lead electro-cardiography, pulmonary function testing, and chest radiography (Jiang *et al.* 2006).

The data collection for the GBCS is facilitated through a strategic alliance with a local welfare association Guangzhou Zunlao Kangle Xiehui (The Guangzhou Health and Happiness Association for the Respectable Elders, GHHARE). The GHHARE was set up by a retiree, Xie Yue, in 1994, and is a welfare association with 'Chinese characteristic'. Namely, it is a privately run organisation strongly associated with the municipal government. Since the Economic Reforms, the Chinese government has gradually withdrawn from the social welfare institutions and let the market or the non-government organisations (NGOs) provide the needed programs. The GHHARE fills the gap of welfare for the elders by providing various kinds of store discounts and purchased services to its members. Anyone 50 years of age or older can join the association for a monthly fee of 4 RMB (about US$0.5) (Jiang *et al.* 2006). By the end of 2006, the GHHARE had branches throughout Guangzhou and has 60,000 members in total. In addition, the GHHARE also serves the state as a mechanism of mass mobilisation. On 17 October 2003, the GHHARE organised a hiking event to celebrate the launch of the manned spaceship *Shenzhou V*, which was described in the official media as 'China's self-designed manned spaceship'.[2] About 3000 members of the association sang and danced with a big red banner dedicated to this successful launch. The *Shenzhou V* spacecraft carried the astronaut Yang Liewei and made China the third nation to send a human into space, 40 years after the Soviet Union and the United States. The event thus became a matter of national prestige. In March 2004, the GHHARE once again called its members to sign up to support Guangzhou's bid for the 2010 Asian Games. The mayor of Guangzhou had officially submitted the application to the Olympic Council of Asia (OCA).[3] In February and March 2004, the OCA sent delegations to inspect facilities in all candidate cities. One of the criteria under review was the citizens' support for the city to host the Asian Games.

Around the same time of the Asian Games application, the GHHARE also helped the GBCS biobank program to collect blood samples. This method of data collection was observed and questioned by a journalist of *Nanfang City News*, a liberal newspaper in southern China.[4] The journalist reported that the GHHARE arranged for forty-two members of a local branch *Summer Grass* (*Xiamao*) to receive a medical examination at the Guangzhou Number 12 Hospital on 9 March 2004. The reporter interviewed ten of them while they were holding questionnaires waiting in line for the interview or examination. The reporter found out that most of the subjects were illiterate, and could not comprehend the questionnaires. Thus, the researchers had to read the questions for the subjects in order to elicit answers. The reporter then asked the subjects whether they knew this free medical check-up was actually for a research project. An old woman replied with confusion, 'What is research (*keyan*)? I have no idea'. All interviewees said that they did not know the medical examination was for a research project. They were interested because the director of the GHHARE branch told them that this was a 'deluxe medical exam free of charge' (*mianfei gaoji tijian*). Thus, these senior people were very motivated to attend; they called cabs and split the fares to come to the hospital. The Deputy Secretary-General of the GHHARE, who was in charge of GBCS subject recruitment, assured the reporter that directors of all branches knew the nature of this research. He said perhaps the term 'deluxe medical exam free of charge' was solely used to attract older members. Yet when the reporter asked the director of the *Summer Grass* whether she knew the physical examination was for research, she shook her head, and said she just knew that it was a free medical check-up. The director of another branch, interviewed by the reporter, replied that he knew the details of the GBCS program. He also suggested that it was difficult to explain the research to some old people, and thus they sometimes had to use 'physical exam' to communicate with them. Overall, it seems that the elder residents in Guangzhou did not know about biobanking despite the fact that they were the major donors of the samples.[5]

Traditionally, Chinese are less willing to give blood because blood donation was associated with loss of *sheng qi* (life energy) (Thomas *et al.* 2006). How about young urban residents? Would they be willing to donate their blood if they knew the nature of the research? The reporter randomly sampled fifteen people on the street and asked their opinions. Most of them replied positively because they considered that such research would ultimately help prevent and control diseases. After reading the news report, some readers called the *Nanfang City News* to ask whether they could donate blood to the GBCS. Some other readers called the *Nanfang City News* to inquire whether this international collaboration posed any risk of 'losing genetic resources' (*jiyin liushi*).[6]

In fact, such risks of 'losing genetic resources' forced the GBCS to set up its biobank within China. In 1998, when Tai-Hing Lam first proposed to Chaoqiang Jiang the idea of launching a biobank, they planned to send all samples to England for storage. According to Jiang, it was very important to properly store biological samples. At that time only a few countries including the USA, Iceland, and the UK had the needed facilities and technicians for managing biobanks. Therefore, in 1999, Lam and Jiang visited the CTSU at Oxford University to seek a possible storage solution. The CTSU promised them that it would keep the blood samples for the project. Yet because of the Chinese legislation 'Interim Measures for the Administration of Human Genetic Resources' (see Appendix) Lam and Jiang could not export genetic samples to foreign countries. To avoid losing genetic resources, blood samples of DNA were required to remain within Chinese borders.[7] Therefore, the GBSC had to set up the biobank at the Guangzhou Number 12 Hospital. In 2003, Jiang *et al.* spent almost half of their initial funding on the facilities. For storing the biological samples, they renovated the basement of the Guangzhou Number 12 Hospital, and imported two MVE 1830 HE freezers,[8] standard equipment for keeping such samples.

In China, the risk of biopiracy by foreigners is more visible than the risk of infringing privacy. Why does there exist such a different understanding of risks in China and in Europe? Ulrich Beck argues that risks are ultimately both real and unreal because many risks completely evade the human capacity of direct perception (Beck 1992). Instead, risks are perceived through *arguments*. Often, there are competing claims in defining risks, and building a consensus on a risk involves convincing others with arguments. When Beck analyses risks, he has global environmental pollution in mind, and in such a case risks are the probabilities of physical harm due to given technological or other processes. In the case of biobanking, we may broaden the concept of risks to, say, the infringement of privacy associated with a genetic database, given that the infringement of privacy as a risk also needs to be perceived by discourses. If risks can only be established by argument, then we may ask the following questions: Who are entitled to define risks, and for whom? What kinds of risks are visible as they are prioritised in the public discourse? What kinds of risks are downplayed and invisible?

Note that in the discussions on the risk of biobanking, the Chinese public rarely brought up the issues of 'privacy'. Does that mean Chinese citizens do not defend their own individual rights? Many China scholars observe that there has been a rise of individuality in post-Mao China across generations, and this tendency is prevalent in both rural and urban settings (Farrer 2002; Liu 2000; Yan 2003). Yan Yunxiang investigated the changes in people's private lives in rural China from 1949 to 1999 and found a rise of private space for privacy in the household. Such space metaphorically indicates 'territories of the self' in the family (Yan 2003:135). Meanwhile, young people more freely express their opinions and values, and have more autonomy in their courtship and marriage decisions. In the public domain, more and more public protests about problems related to housing, tax, or education have been reported in recent years. So it is not that Chinese citizens do not defend their own individual rights, but the ideas about when a right is infringed upon are different.

The responses towards biobanks also indicate a specific understanding of risk in relation to science. In post-Mao China, science has been perceived as a 'productive force' for economic development, and an ultimate solution for all kinds of problems. The thoughts associated with science have always been positive. By contrast, in the West, science has been a new target for suspicion partly because of the darker consequences of scientific and industrial development in creating environmental risks and hazards. Ulrich Beck points out that human history is not an ideal progressive development. The first modernisation is to de-mythify religion through rationalisation. Yet science now has become a 'higher superstition', creating environmental pollution, among other harms. Thus, the second modernisation is to de-mythify science (Beck 1992). Science is also under suspicion owing to the close relation between knowledge and power. As Michel Foucault points out, the governmentality of the modern state is based on rationality and expertise. Power not only occurs in the format of force, but also in other formats seemingly irrelevant to power. For example, medical doctors are double agents because they not only cure patients, but also serve to maintain the social order, and draw the boundary between normal and abnormal in society.

By contrast, such a distrust of science rarely appears in social discourses in China. The contrast between China and the West toward science can be illustrated by their response toward pollution. China is now facing increasingly serious problems of pollution. Yet the Chinese perceive pollution as the outcome of inappropriate management or corruption, something *external* to science. Under this perception, pollution can eventually be brought under control by means of rational, science-based policies and effective execution, i.e. by being *more* scientific. In contrast, the Western environmentalists often stress the limit of science in preventing and reducing environmental problems. As Rachel Carson wrote in *Silent Spring*, pollution cannot be foreseen and we should not leave the planning and policies of controlling pollution only in the hands of scientists and experts (Carson 2002).

In China, many scientific policies are based on this cultural perception of science, and thus appear radical and extreme to foreign observers. Consider the one-child policy. The one-child policy was based on the projection of population growth calculated by missile scientist Song Jian. At that time, Chinese social scientists were against this one-child policy, but the unavailability of super-computers to calculate the growth of the Chinese population did not enable them to substantiate their argument. In the end, the government decided to choose a population control policy based on a 'scientific' approach (Greenhalgh 2005). The promotion of genetically engineered crops and the Three Gorges Dam are other indications of the positive attitude of the Chinese towards science. These scientific or technological projects represent active choices that prefer economic development to any potential risks of technology.

In the West, along with the distrust of science, there has been a demand for transparency in science. Transparency means different practices in different societies. In the USA, transparency means to make the *procedures and results* of scientific policies available to the public. In Germany, transparency of scientific policies means that the government or the policy-makers need to make the *rationales* of decision clear to the public. All these kinds of transparency practices invite the public to participate in the debates on science. These measures aim to ensure that the decisions about science are not monopolised by experts. By contrast, in China, decisions concerning science are not transparent to the public, which parallels the public's strong trust of science. Science policies are usually left in the hands of the Chinese government and scientists alone.

In short, one dominant discourse about biobanking in China is the risk of 'losing genetic resources' (*jiyin liushi*), which has been mentioned by government officers, scientists, the general public, and journalists. Such a popular perception reveals the lack of international norms on biobanking. It also indicates distrust at another level, that of international collaboration. In other words, the risk of biopiracy cannot be analysed solely within the framework of Chinese cultural or social factors. Rather, it needs to be understood in terms of the 'genetic resources' from the perspective of global research and development. To do so, we have to trace the very meaning and genealogy of 'genetic resources' in China.

Discourse on national resource

No resource is purely 'natural' or given. A windmill converts wind into mechanical energy; a dam transforms water into power. Definition and uses of resources are mediated by knowledge and technologies. Furthermore, as science and technology develop, new kinds of resources emerge. In recent developments in life science, the aborted foetus has changed from waste to a valuable resource because scientists can extract stem cells from an aborted foetus for their research. In other words, in the wonder of regenerative medicine, stem cell research serves as a translator to transform the foetus from valueless into valuable. Similarly, the concept of 'genetic resources' began to circulate in the early 1990, when the new discipline of genomics allowed scientists to apply more powerful means to study the genetics of different species. As a result, tumour tissues which used to be considered waste or archive materials are now valuable resources for genetic analysis.

At the one end of genetic resources, there lies science and knowledge; at the other end, there lies biodiversity, environment conservation, agriculture, and the pharmaceutical industry. To mobilise their genetic resources for economic development or environment conservation, both rich and poor nations raced to initiate multiple programs. In 1992, the USA established the National Genetic Resources Program (NGRP) in order to 'acquire, characterize, preserve, document, and distribute to scientists, germplasm of all lifeforms important for food and agricultural production'.[9] Also in 1992, the UN Convention on Biological Diversity was enacted to promote 'the conservation of biological diversity, the sustainable use of its components, and the fair and equitable sharing of benefits arising out of the utilization of genetic resources'.[10] Riding on the trend of acquiring human genetic resources, more and more biobanks are being set up or expanded.

From the beginning, the discourses and practices of 'genetic resources' have been embedded in the complicated political economy of global research and development. Developing countries are perceived as 'gene-rich but cash-poor', and attract scientists and the pharmaceutical industry from the developed countries. Since the 1980s and 1990s, the pharmaceutical industry has simultaneously adopted multiple approaches to utilise genetic resources in drug invention. On the one hand, the industry searches for effective plant compounds from different healing traditions for drug development (Hayden 2003). On the other hand, after the international Human Genome Project was set up in 1990, there has been a global gold rush of human genes. Scientists in the technologically advanced countries not only established programs to acquire human genetic resources within their own countries, but also instigated international projects to collect additional gene samples from foreign countries (Rabinow 1999b).

At that time, the Chinese population appeared in the Western public media and scientific journals as a storehouse of rare genetic resource; China comprises fifty-six ethnicities, and it also has relatively isolated and homogeneous populations in its interior. The very 'marked rural and urban contrasts' in China are considered the best illustration of the health effects of rural–urban migration (Holden 1996). This perception drove Western scientists to China collect blood samples of genes. For the Western researchers, China not only provided rare genetic data that could not be obtained elsewhere, but also offered a 'low-cost research venue' for genetics research.

In China, the 'Anhui genes event' (*Anhui jiyin shijian*) occurring in the late 1990s was critical in formulating public discussion of the risks of biopiracy.

On 19 July 1996 *Science* magazine reported a large-scale sample collection (Holden 1996). The director of a new program of population genetics at Harvard University, Xiping Xu, would collect 200 million blood samples of DNA in remote areas of Anhui Province through collaboration with six Chinese medical centres. The program aimed to study the 'diseases of civilization', including diabetes, obesity, hypertension, early-onset heart disease, schizophrenia, and some infectious diseases. Although Xu grew up in Anhui, the Chinese mass media still described Xu's research as a case of exploitation by foreigners because Xu was affiliated with both Harvard University and the Millennium Pharmaceutical Company. Xu's research would ultimately benefit these American institutes but not Chinese ones.

This report first provoked a series of events initiated by several groups of Chinese scientists, and then set the stage for China to enact the legislation 'Interim Measures for the Administration of Human Genetic Resources' and launch several scientific institutes to harness their own human genetic resources. The first public response was by Henry Yang, the then Chinese Human Genome Project Secretary-General, on 4 February 1997. In an interview entitled 'Battle of the Century' (*Shiji Zhi Zhan*) in the *China Youth Daily*, Yang concluded the interview: 'if we let others steal "China's valuable gene resources, we will pay a heavy price ... If we miss this opportunity, we will be ashamed of confronting our ancestors, our children and our grandchildren!"' (Anonymous 1997). One month later, an article in the *Journal of the Chinese Academy of Medical Sciences* (*Zhongguo Yixue Kexueyuan Yuanbao*) warned that China faced the prospect that the genes of its populations might become priceless resources for foreign pharmaceutical companies. In April, another Chinese scientist, Qiang, published an article, 'Protect Our Country's Human Genome Resource', in the *Healthy Daily* newspaper to call for legislation to protect China's genetic resources. Qiang warned that US companies are exploiting Chinese genetic resources without providing adequate protection or compensation (Anonymous 1997). Such declarations from scientists created tension between the USA and China. In April 1997, the US Embassy in Beijing therefore issued a report entitled 'Alarm at US Companies Draining China's Gene Pool' (Anonymous 1997).

The next stage of events further captured the attention of political leaders in China. On 15 July 1997, one of the founders of genetics in China, Tan Jiazhen, submitted a letter to then President Jiang Zemin suggesting that Chinese genetic resources had to be protected. One week later, Jiang Zemin replied in an official memorandum entitled 'We Have to Cherish Our Genetic Heritage' (Li and Wang 1997). Shortly thereafter, the Chinese government established the regulations for Chinese human genetic resources, 'Interim Measures for the Administration of Human Genetic Resources'. By these regulations, any foreign scientists who want to obtain samples of genes from China have to collaborate with Chinese scientists. The 'Interim Measures' do not explicitly forbid the exportation of genetic samples to foreign countries. In actual practice, however, Chinese government officials give a more strict interpretation of the regulations, resulting in a de facto rule that genetic resources cannot be exported and have to remain within national borders.

By setting up regulations regarding the Chinese genetic heritage in 1998, Chinese scientists became the gatekeepers of these genetic materials. On the one hand, they used this phrase with a nationalist sentiment to urge the state to protect Chinese DNA because it was 'ours', the common heritage of all Chinese. On the other hand, by voicing the need to protect Chinese DNA from exploitation by foreigners, Chinese scientists obtained a monopoly on Chinese DNA.

The same group of Chinese scientists who first advocated the regulations in 1997 later lobbied the Chinese government to set up three genome centres in 1998 and 1999 in order to protect 'genetic heritage'. They themselves became directors of these three genome centres, and claimed genome centres as important sites for appropriate genetic resources in China.

Conclusion: biobanking, risk, bounded and non-bounded resources

This chapter has examined some of the risk discourses surrounding the establishment of biobanks and the corresponding controversies in China. Whereas Europeans and North Americans are more concerned with the biobanks' potential intrusion into individual privacy, the Chinese public worries more about foreign countries stealing China's genetic resources via internationally collaborated biobanks. Using Ulrich Beck's analysis of risk, the author argues that this difference in the perception of the dangers related to biobanks results from distinct ways of constructing the risk discourse, which reflects the West's and China's different value judgements of science. In the Chinese context, therefore, the issues of risk in biobanking are closely related to the defence of population genes as national resources against foreign 'biopiracy'.

This awareness of 'biopiracy' led the Chinese government and scientists to enact regulations and develop research institutions to preserve, utilise, and protect the 'genetic resources' of the country. Yet the very notion of genetic resources is ambiguous as it connotes a pair of seemingly opposite attributes: genetic resources are both bounded and non-bounded. On the one hand, genetic resources, like other natural resources, are attached to specific places. In the two large-scale Chinese biobanks, the GBSC and the KSCDC, the geographic factors represent degree of urbanisation, lifestyles, culture, and environment variables. Thus, biological data derived from overseas Chinese cannot replace that from Chinese living in China. Given that a prospective cohort study is to investigate the interplay between genetic and environment factors in common chronic diseases, whether a biobank connotes regional or environment information determines the value of this biobank. On the other hand, while the traditional natural resources consist of the substances retrieved from animals, plants, and mines, the essence of the genetic resources – and hence the true value of the biobanks that store these resources – lies not in the fabric of blood and tissue samples. Rather, the essence lies in the *information* carried by the material samples. The genomic information can be transported anywhere without losing or changing its content, at least in principle; similar in a way to what Bruno Latour has called 'immobile movable'. From this perspective, it may be off the point, if not totally irrelevant, for the Chinese government and scientists to guard the blood and tissue samples from being transported abroad. What they really need to do is to prevent the outflow of the genomic data as well. And the latter cannot be achieved simply by locating the biobanks within the national borders, for the genetic resources as information are non-bounded.

Both the bounded and non-bounded characteristics are essential for understanding further the notion of the genetic resources. Whereas the bounded attributes point to the complex relationship between genes and environment, the non-bounded attributes remind us of the informational dimension of the new genomic science. In the story of biobanking in China, moreover, all these intellectual factors are played out in a mesh of globalisation, nationalism, and renewed meanings of natural resources.

Appendix

Interim Measures for the Administration of Human Genetic Resources Promulgated by the General Office of the State Council upon the approval of the State Council on 10 June 1998
The Ministry of Science and Technology
and
The Ministry of Public Health
The People's Republic of China

Chapter One: General provisions

Article 1

The Interim Measures for the Administration of Human Genetic Resources (hereinafter referred to as the Measures) are enacted for the purpose of efficiently protecting and rationally utilizing human genetic resources in the People's Republic of China, strengthening the research and development of human genes and promoting international cooperation and exchange on the basis of equality and mutual benefits.

Article 2

The term 'human genetic resources' in the Measures refers to the genetic materials such as human organs, tissues, cells, blood specimens, preparations of any types or recombinant DNA constructs, which contain human genome, genes or gene products as well as to the information related to such genetic materials.

Article 3

Whoever is involved in such activities in China as sampling, collecting, researching, developing, trading or exporting human genetic resources or taking such resources outside the territory of the People's Republic of China shall abide by the Measures.

Article 4

The State adopts a reporting and registration system on important pedigrees and genetic resources in specified regions. Any institution or individual who discovers or holds important pedigrees and genetic resources in the specified regions shall immediately report to the relevant departments. No institution or individual may sample, collect, trade, export human genetic resources or take them outside the territory of the People's Republic of China, or provide them to other countries in any form without permission.

Article 5

Where the human genetic resources and the relevant information or data are classified as State scientific or technological secrets, the Rules for the Protection of State Secrets in Science and Technology shall be observed.

Chapter Two: Administration

Article 6

The State adopts a unified review-and-approval system regulated at different levels over human genetic resources.

Article 7

The administrative department of science and technology and the administrative department of public health under the State Council shall be jointly in charge of the administration of human genetic resources of China on a national scale and shall jointly establish the Human Genetic Resources Administration of China (hereinafter referred to as the HGRAC) to carry out routine duties.

Article 8

The HGRAC is initially placed under the administrative department of science and technology under the State Council. Under the direction of the administrative department of science and technology and the administrative department of public health under the State Council, the HGRAC shall perform the following responsibilities.

1   To draft the relevant rules and implementary documents, promulgate such rules for entering into force upon approval and ensure the enforcement of the Measures through co-ordination and supervision;
2   to be in charge of the registration and administration of the important pedigrees and genetic resources in the specified regions;
3   to review and examine international collaborative projects involving human genetic resources in China;
4   to review and approve applications for exportation of human genetic resources, and thereafter to issue an Export Permit for Human Genetic Materials (hereinafter referred to as the Export Permit; and
5   other duties related to the administration of human genetic resources in China.

Article 9

An expert panel shall be formed by the HGRAC to participate in formulating research plans, assist in reviewing and examining international collaborative projects, undertake relevant technical appraisal and provide professional consultation.

Article 10

The administrative department of science and technology and the administrative department of public health at the level of provinces, autonomous regions or municipalities directly under the Central Government (hereinafter referred to as the local administrative departments), shall be in charge of the administration of human genetic resources in their own respective regions.

The relevant departments under the State Council shall be in charge of the administration of human genetic resources within their own administrative domains.

Chapter Three: Application, examination and approval

Article 11

Where human genetic resources in China are involved in any international collaborative project, the Chinese collaborating party shall be responsible for going through the due formalities of application for approval. Institutions directly under the Central Government shall apply to the relevant administrative department under the State Council and local institutions or institutions without a specific supervisory department shall apply to the local administrative departments, and upon receiving the approval of the relevant departments, the Chinese collaborating party shall apply to the HGRAC for examination and approval prior to entering into an official contract.

Relevant departments under the State Council and local administrative departments, in reviewing any application for international collaborative projects, shall consult the relevant local administrative departments of the region where human genetic materials are to be collected.

Any international collaborative project which has been carried out but has not been completed prior to the implementation of the Measures shall follow up application for examination and approval as a makeup in accordance with the Measures.

Article 12

An application form shall be filled in and the following documents be included in going through the applying and approving formalities of an international collaborative project involving human genetic resources of China:

1 Informed consent form of the donor of the human genetic materials and/or his (her) legal representatives;
2 draft contract; and
3 other documents required by the examining and approving departments.

Article 13

No application submitted in accordance with Article 12 of the Measures shall be approved if any of the following circumstances is involved:

1 Where the project lacks a precise objective or purpose;
2 where the foreign partner fails to possess adequate research capability or advantage in the research and development;
3 where the Chinese partner fails to possess the required basis and conditions for the collaborative research;
4 where the proportioning of ownership and share of intellectual property right is unfair or unclear;
5 where the collaboration exceeds the proper scope or time limit;
6 where informed consent forms of the donor of the human genetic materials and/or his (her) legal representatives are lacking; or
7 where the relevant State laws or rules and regulations are violated.

Article 14

Any export of important human genetic resources outside the territory of the People's Republic of China or provision of such human genetic resources to foreign institutions or individuals shall be under strict control.

Where, in an international collaborative project which has already been examined and approved, the plan is made for the report of human genetic materials outside the territory of the People's Republic of China, the application form shall be filled in and the Export Permit issued by the HGRAC.

Where, under special circumstances, temporary provision of human genetic materials to other countries is needed, the application form shall be filled in and submitted to the HGRAC for approval upon the examination and approval of the local administrative departments or the relevant administrative departments under the State Council, and then the Export Permit shall be issued upon the approval of the HGRAC.

Article 15

The HGRAC shall handle the applications for international collaborative projects and export of human genetic materials outside the territory of the People's Republic of China every quarter. Approval shall be granted if the applications meet the requirements stipulated in the Measures, the Export Permit issued and the corresponding code stipulated in the Harmonized Commodity Description and Coding System noted; no approval shall be granted if the applications fail to meet the requirements stipulated in the Measures: and such applications as fail to provide adequate documents shall be returned for revision and re-applying shall be permitted upon revision.

Article 16

All human genetic materials to be exported by means of hand carrying, mailing and transporting should be truthfully declared to the Chinese Customs. The Chinese Customs will give clearance to those accompanied by the Export Permit issued by the HGRAC.

Chapter Four: Intellectual property right

Article 17

The Chinese research and development institution shall have the priority to access information about the human genetic resources within the territory of the People's Republic of China, particularly the important pedigrees and genetic resources in the specified regions and the relevant data, information and specimens and any transfer of such human genetic resources to other institutions shall be prohibited without permission. No foreign collaborating institution or individual that has access to the above mentioned information may publicise, publish, apply for patent right or disclose it by any other means without permission.

Article 18

International collaborative projects involving human genetic resources shall follow the principles of mutual benefits, credit and trust, joint participation, and share of achievements; all rights and obligations of each party should be explicated in order to fully and effectively protect their own respective intellectual property rights.

Article 19

In a collaborative research and development of human genetic resources of China between any Chinese and foreign institutions, intellectual property right therefore shall be disposed according to the following principles:

1   Patent shall be jointly applied for by both parties and the consequent patent right shall be owned by both parties if an achievement resulted from the collaboration is patentable. Either party has the right to implement such patent separately or jointly in its own country in accordance with the contract; however, the transfer of such patent to any third party or the permission of authorizing any third party to implement such patent shall be carried out upon agreement of both parties, and the benefits obtained thereof shall be shared in accordance with their respective contributions.
2   The right of utilizing, transferring and sharing any other scientific achievement resulting from the collaboration shall be specified in the collaborative contract or agreement signed by both parties. Both parties are equally entitled to make use of the achievement which is not specified in the contract or agreement; however, the transfer of such achievement to any third party shall be carried out upon agreement of both parties, and the benefits obtained thereof shall be shared in accordance with their respective contributions.

Chapter Five: Rewards and penalties

Article 20

Any institution or individual that discovers and reports important human genetic resources shall be praised and rewarded: whoever exposes illegal activities shall be rewarded and protected.

Article 21

If any Chinese institution or individual, in violation of the provisions stipulated in the Measures, exports the human genetic materials without authorization by hand carrying, mailing, or transporting, the human genetic materials shall be confiscated by the Chinese Customs and the institution or individual shall be punished ranging from administrative sanctions to prosecution by the judicial department according to the seriousness of the circumstances. If anyone, in violation of the provisions stipulated in the Measures, provides human genetic materials to foreign institutions or individuals without permission, the human genetic materials shall be confiscated and the institution or individual shall be fined; if the circumstances are serious, he shall be investigated for legal responsibility according to Chinese law.

Article 22

If anyone exports the human genetic resources outside China by hand carrying, mailing, or transporting without authorization, the human genetic materials shall be confiscated by the Chinese Customs and he shall be punished or put under the prosecution of the judicial department according to the seriousness of the circumstances.

Article 23

Any staff member of the administrative department or expert engaging in the examination shall have the duty to keep technological secrets for the applicants. Whoever causes the exposure of technological secrets or loss of the human genetic resources of China due to negligence in his duty or malpractice for personal gains shall be imposed with a punishment ranging from administrative sanctions to being investigated for legal responsibility.

Chapter Six: Supplementary provisions

Article 24

The military forces may formulate their own specific implementory rules in accordance with the provisions stipulated in the Measures and submit to HGRAC for record. The armed police force shall observe the Measures.

Article 25

The administrative departments of science and technology and the administrative department of public health under the State Council shall be responsible for the interpretation of the Measures.

Article 26

The Measures shall enter into effect on the date of promulgation.

Source: National Lab of Medical Genetics of China, http://www.ocgr.org/Cn/laws/042315022302.aspx

Notes

1 The role of the CTSU in the KSCDC project is not very clear in the publication by KSCDC researchers. It is also not clear where the KSCDC stores its blood samples.
2 Yangcheng Evening News, 17 October 2003.
3 Guangzhou Television Station News, 31 March 2004.
4 Nanfang City News, 10 March 2004.
5 Nanfang City News, 10 March 2004.
6 Nanfang City News, 10 March 2004.
7 Nanfang City News, 9 March 2004.
8 Nanfang City News, 9 March 2004.
9 Agricultural Research Service at the United States Department of Agriculture, http://www.ars-grin.gov/
10 UN Convention on Biological Diversity, http://www.biodiv.org/convention/articles.shtml?a=cbd-01

References

Anonymous (1997) 'China Elevates Scientists to Party Posts', *Science* 277(5334): 1915d.

Beck, Ulrich (1992) *Risk Society: Towards a New Modernity* (transl. M. Ritter). London: Sage Publications.

Cambon-Thomsen, Anne (2004) 'The Social and Ethical Issues of Post-Genomic Human Biobanks', *Nature Reviews Genetics* 5(11):866–73.

Carson, Rachel (2002) *Silent Spring* (40th edn). Boston: Houghton Mifflin.

Chen Zhengming, Liming Lee, Junshi Chen, Rory Collins, Fan Wu, Yu Guo, Pamela Linksted, and Richard Peto (2005) 'Cohort Profile: The Kadoorie Study of Chronic Disease in China (KSCDC)', *International Journal of Epidemiology* 34:1243–9.

Farrer, James (2002) *Opening Up: Youth Sex Culture and Market Reform in Shanghai*. Chicago: University of Chicago Press.

Greenhalgh, Susan (2005) 'Missile Science, Population Science: The Origins of China's One-Child Policy', *The China Quarterly* 182:253–76.

Hayden, Cori (2003) *When Nature Goes Public: The Making and Unmaking of Bioprospecting in Mexico*. Princeton, NJ: Princeton University Press.

Hirtzlin, Isabelle, Christine Dubreuil, Nathalie Preaubert, Jenny Duchier, Brigitte Jansen, Jurgen Simon, Paula Lobato de Faria, Anna Perez-Lezaun, Bert Visser, Garrath D. Williams, and Anne Cambon-Thomsen (2003) 'An Empirical Survey on Biobanking of Human Genetic Material and Data in Six EU Countries', *European Journal of Human Genetics* 11:475–88.

Holden, Constance (1996) 'Harvard and China Probe Disease Genes', *Science* 273:315.

Jiang Chaoqiang, G. Neil Thomas, Tai Hing Lam, C. Mary Schooling, Weisen Zhang, Xiangqian Lao, Peymane Adab, Bin Liu, Gabriel M. Leung, and Kar Keung Cheng (2006) 'Cohort Profile: The Guangzhou Biobank Cohort Study, Guangzhou-Hong Kong–Birmingham collaboration', *International Journal of Epidemiology* 35:844–52.

Keller, Evelyn Fox (1992) 'Nature, Nurture, and the Human Genome Project'. In *The Code of Codes: Scientific and Social Issues in the Human Genome Project* (D. J. Kelves and L. Hood, Eds). Cambridge, MA: Harvard University Press, 281–99.

Lao, X.Q., G.N. Thomas, C.Q. Jiang, W.S. Zhang, P. Yin, M. Schooling, M. Heys, G. M. Leung, P. Adab, K.K. Cheng, and T.H. Lam (2006) 'Parity and the Metabolic

Syndrome in Older Chinese Women: The Guangzhou Biobank Cohort Study', *Clinical Endocrinology* 65(4):460–9.

Lewontin, Richard C. (1992) *Biology as Ideology: The Doctrine of DNA*. New York: Harper Perennial.

Li, Hui, and Jue Wang (1997) 'Backlash Disrupts China Exchanges', *Science* 278(5337):376–7.

Liu, Xin (2000) *In One's Own Shadow: An Ethnographic Account of the Condition of Post-reform Rural China*. Berkeley, CA: University of California Press.

Rabinow, Paul (1999a) 'Artificiality and Enlightenment: From Sociobiology to Biosociality'. In: *The Science Studies Reader* (M. Biagioli, Ed.) New York: Routledge, 407–16.

Rabinow, Paul (1999b) *French DNA: Trouble in Purgatory*. Chicago: University of Chicago Press.

Thomas, G.N., C.Q. Jiang, Sarah M. McGhee, Wei Sen Zhang, Xiang Qian Lao, Mary Schooling, Peymane Adab, Tai Hing Lam, and Kar Keung Cheng (2006) 'Association of Vascular Risk Factors with Increasing Glycemia Even in Mormoglycemic Subjects in an Older Chinese Population: The Guangzhou Biobank Cohort Study', *Metabolism: Clinical and Experimental* 55: 1035–41.

Yan, Yunxiang (2003) *Private Life Under Socialism: Love, Intimacy, and Family Change in a Chinese Village: 1949–1999*, Stanford, CT: Stanford University Press.

10 Bioethical regulation and human genetic databases in Mainland China

A national survey among scientists and regulators on consent issues and benefit-sharing

Xinqing Zhang

Introduction

China is the only developing country that joined the international Human Genome Research Project. Its so-called '1% Project', sequencing 1 per cent of the human genome, drew much attention from not only the domestic biomedical community, but also the highest policymakers in China. To facilitate this effort, the Chinese Academy of Science (CAS) set up the Beijing Institute of Genomics, led by Professor Yang Huanming. So as to catch up with the latest developments in this highly competitive field, China's Ministry of Science and Technology (MoST) in 1998 established its National Human Genome Research Centers located in Beijing and Shanghai. These two national centres have announced the establishment of several databases focusing on common diseases such as hypertension. For instance, the database for Genomic Polymorphisms of Chinese Ethnic Groups (GPCEG) contains denomination and basic information, such as data of genomic polymorphism, cell lines and references of Han and another fifty-five national minorities.[1]

Since the late 1990s, the number of genetic databases has increased rapidly in Mainland China. Projects related to the collection of large Chinese genetic samples have also drawn attention.[2] In fact, the collection, processing, use, and storing of human genetic samples may involve actual and potential ethical problems, reflected in sustained and intense international debate. As a reaction, in 1998, the MoST and the Ministry of Health (MoH) issued the 'Interim Measures for the Administration of Human Genetic Resources' (Interim Measures hereafter). In order to improve its implementation, the two ministries set up a joint office in charge of reviewing relevant protocols. Although the Interim Measures play an important role in regulating the foreign involvement of international genetic study projects, they do not constitute a clear legal framework or ethical guidelines for genetic research involving large numbers of human genetic samples in China. This omission implies a great imbalance between the rapid increase of genetic data collecting and use and the neglect of ethical guidelines and regulation. This is also the case in relation to consent issues.

This chapter focuses on the attitudes and perceptions of Chinese genetic scientists of ethical issues in the process of the creation of genetic databases. A nationwide mail survey was conducted on human genetic bases in various Chinese cities, such as Shanghai and Beijing, to identify whether the Chinese research communities recognised fundamental ethical topics, such as informed consent, privacy, ownership, commercial conflict of interests, and ethical review mechanisms. A total of 300 Chinese scientists and Ethical Review Committee (ERC) members returned the valid questionnaires, with a response rate of 77 per cent.

Survey method

Until now, little knowledge has been available on the attitudes and perceptions of Chinese biomedical communities regarding the creation of genetic databases. A main objective of the survey was to find out whether Chinese research communities identified the main ethical issues that have been heatedly discussed over years on an international level. Secondly, it aimed to explore their attitudes towards the ethical issues that arose with developing genetic databases in China. It also wanted to know what interventions and strategies the research community thinks are best so as to avoid these ethical constraints.

The study hopes to provide some insight into Chinese attitudes towards ethical issues regarding genetic databases, to identify the central friction points, and to encourage further discussion. As an elite, professional group, the target population of this survey cannot be representative of the Chinese population as a whole. The discussion, therefore, should be regarded as a preliminary discussion to broader debate. We selected our respondents using the following criteria:

1   Principle Investigators (PIs) that are faculty of Chinese academic institutes involving human genetic databases in China;
2   their research is closely relevant to human genetic databases in China;
3   PIs and their research teams have published their relative findings in PubMed or other top domestic journals since 1998;
4   there are ERCs or equivalents in those institutes to which the PIs belong.

The target institutions of this nationwide mail survey included institutions and affiliated hospitals relevant to the National Human Genome Research Centers (in Beijing and Shanghai) and several other provincial genetic databases.[3]

The survey was conducted in July–October, 2005. A total number of 300 out of 390 valid questionnaires were returned. The response rate was 77 per cent.

The anonymous questionnaire included a section on socio-demographic data, followed by questions on ethical issues. The respondents included 166 males and 134 females: 52.4 per cent were between 31 and 50; 29 per cent came from Beijing, 37.1 per cent from Shanghai, and others from medical institutions and hospitals in other provinces; 37.8 per cent had studied or worked abroad; and 64 per cent of them held PhDs or Master degrees. The majority of them had majored in the field of biomedicine. In the sampled population, 74 per cent used human genetic samples directly in their work, and 55 per cent collected samples by themselves. Fifty-three per cent claimed that they had been involved in databases for more than one year. Fifteen per cent said that they had reviewed protocols related to genetic data collection and usage. Respondents who are not ERC numbers but reviewed the protocols as well usually were engaged in scientific administration offices and/or department of medical affairs.

As for their attitude to ethical issues in general, 53 per cent said they have considered the ethical issues in the process of establishing databases, 31 per cent said it is urgent to identify and solve these issues, 12 per cent said it was little to do with their current research, and the other 4 per cent said that it is unnecessary to talk about ethical issues now. Of the respondents, 54.7 per cent of them did not participate in any kind of research ethics training courses, 34 per cent took part in short training courses, and 11.3 per cent had studied research ethics systematically.

It can be concluded that in general our respondents knew little about the international ethical discussion. More than half of the respondents (55.3 per cent) reported that they had paid no attention to the ethical debates in Iceland, the UK, Estonia, and Tonga. About one-fifth of respondents (20.4 per cent) reported that they had an idea about the ethical debate in Iceland, and 32.4 per cent had heard of UK Biobank.

Attitudes about consent policy

In the international debate on genetic databases, probably the most hotly debated issue is that of consent. Actually, in China a similar situation exists with regard to the conduct of clinical trials. But considering the rapid growth of the number of genetic databases in China, not many articles have been written and little public discussion taken place about consent issues in China. In the international context of genetic databases a number of specific issues have arisen that have led to a renewed debate about the limits of current consent models and whether new, different notions of consent need to be considered. For instance, in the Icelandic Health Sector Database presumed consent was voiced in the process of genetic sample collection.[4] In other cases, such as UK Biobank, individual consent is clearly required. We hoped that our survey would yield information on opinions and attitudes of Chinese scientists and ERC members towards those consent issues arising when developing genetic databases. We also wanted to find out what interventions and strategies the research community recommends.

Different forms of informed consent

Following the Nuremberg Code and Helsinki Declaration, individual informed consent is considered a well-known principle of biomedical research at an international level. However, research on population genetics required a new interpretational framework. The individual and collective characteristics of genetic information provide a totally new dimension in the context of large genetic sample collecting and usage. For, depending on the context and the speaker, the requirement of individual consent may not be absolute, while family or community consent may be reasonable and appropriate in other cases. In the case of UK Biobank, an individual written informed consent is required for participation, while in the case of the Icelandic Healthcare Database the Act on the Biobank allowed the use of clinical medical samples for research using the opt-out approach.[5] As the Icelandic approach raised great controversy at the international level, we decided to explore whether Chinese colleagues were aware that presumed consent in the Icelandic case is one of the causes of its failures.

Table 10.1 lists several kinds of consent and the attitude of the sampling population. In the survey, 69 per cent disagreed that oral informed consent could take the place of a written consent. Fifty per cent agreed that if donors from rural areas are not used to providing a signature, fingerprinting would be a satisfactory alternative; 35 per cent disagreed that it is acceptable to use fingerprinting instead of a personal signature. Although family consent is assumed to be popular in the Chinese research setting, only 24 per cent think it is appropriate, 56 per cent disagree with this, and 20 per cent did not make a choice. The results also indicate that many of the respondents disagreed about the use of various forms of consent in the process of DNA databanking in China.

In fact, informed consent is a relatively new concept in China, introduced in the 1980s. At that time, informed consent was actually used in clinical practice. For instance, it used to be common for a husband to sign a written consent form on behalf of his wife's or his kin's surgery. Furthermore, it was unusual in the late 1980s that a patient gave written informed consent when participating in a clinical trial. Before the 1990s there was a long period during which investigators only obtained oral informed consent from subjects. In some cases, even famous Chinese physicians/investigators just informed human subjects that

Table 10.1 Attitudes about different consent forms

Case: Do you agree with the following statement?	Agree	Disagree	No opinion
a Oral consent could take the place of written consent	20%	69%	11%
b Fingerprints could take the place of a personal signature	50%	35%	15%
c Family consent could take the place of the donor's individual consent	24%	56%	20%

they would receive a kind of innovative therapy, rather than telling them they took part in a form of human experimental research.[6]

Since the late 1990s, the health authorities have begun to take measures to protect research subjects. In 1998, the MoH issued the *Procedures for Ethical Review of Human Biomedical Research*. One year later, China's State Food and Drug Agency (SFDA) carried out its own version of Good Clinical Practice (GCP), which is formulated to ensure that the clinical trial process is standardised, and the rights, benefits and safety of trial subjects is protected.[7] In general, the informed consent forms should be signed and dated by the subject or the subject's legally acceptable representative. Although written individual informed consent in biomedical research is required, different forms of consent may be valid in different situation: according to Chinese GCP, when both the subject and his/her legally acceptable representative are unable to read, oral consent is permitted provided that a witness is present and signs the consent form.[8] Since the spread of GCP, individual consent has been applied and promoted from pioneer institutes to others.

'The right to know' versus 'the right to not know'

Unlike the Estonian Genome Project, in the case of UK Biobank, the voluntary donors will not have the right to receive personal feedback on the genetic information derived from samples. Against this background, the survey asked Chinese scientists if Chinese donors have the right to know the research results pertaining to the health of the donors. We also wanted to find out in what situation Chinese scientists think that investigators should disclose research results to donors, even when it is difficult for many of the donors to understand the complex meanings and implications of those scientific results. The following question is supposed to identify the respondents' attitude about the donors' 'right to know': Should we inform donors of the results that contain important therapeutic information?

If negative genetic information were to be disclosed, it could result in stigma and discrimination of individuals and groups, and it could hamper their employment situation and possibilities, their education possibilities, and their ability to obtain medical insurance. For reasons such as these, special safeguards are needed for confidentiality and privacy. Given the obvious difference in the reactions among the respondents (Table 10.2), the option of whether the research findings should be distributed to research participants should be added to the informed consent form. In 1999, the National Bioethics Advisory Committee (NBAC) recommended the following: donors may obtain the final research results, provided that: 1) the findings are scientifically valid and confirmed; 2) the findings have significant implications for the subject's health concerns; 3) a course of action to ameliorate or treat these concerns is readily available.[9] In this case, the option of informed consent could be that the researcher would let the donors decide whether they would like to know the final results pertaining to their health. Such an option, however, presumes awareness among both

Table 10.2 The respondents' attitudes toward 'the right to know'

Should we inform donors of the results that contain important therapeutic information?	Agree	Disagree	No opinion
a Fully inform the donor of the results pertaining to the donor's health, unless the donor himself gives up the right	51%	25%	24%
b Inform the donor of the basic research findings, unless the donor gives up the right	22%	46%	32%
c It is unnecessary to inform the donor of the results, unless the donor requires them	45%	37%	18%

researchers and donors of the social, political, and economic consequences of having knowledge of one's current health and health prospects.

Re-consent issues

It is very difficult to fully inform donors about the research aims when the future uses of DNA samples/data are unknown. In Tonga, a choice was offered to the participants to consent to the use of their samples and data for multiple research projects or for a limited number of predefined projects.[10] In the UK, according to the protocol, consent will be asked for various analyses, for specified and unspecified biochemical and genetic tests, and for permission to contact participants again at a later date.[11] In China, by contrast, the current regulations, such as the Interim Measures for the Administration of Human Genetic Resources, do not mention re-consent. Therefore, we hoped to gather more data on the attitudes of Chinese scientists about re-consent issues, especially in the context of genetic data collecting, storage, and usage.

In response to the question of whether the DNA samples collected for a particular purpose could be used for other future research without the donor's re-consent, even if the samples are identifiable, we obtained the following reactions: 59.7 per cent of the respondents gave a positive answer, even if the samples could be linked with the original donors; 32.7 per cent gave a negative answer; while 7.7 per cent of the respondents did not choose. These results indicate that the majority of the respondents do not think it is necessary to reaffirm consent every time a significant change to the protocol or to the banking conditions occurs. In view of the different attitudes about re-consent, it may be better to provide donors with consent options on informed consent forms. The donors could be provided with options that specify types and conditions of future research for which their samples may be used.

Arguments for and against family consent

The Human Genome Organisation (HUGO) claims that 'informed consent can be individual, familial, or at the level of communities and populations'.[12] In

China, family consent is considered to play a key role in clinical practice. A possible rationale for this is that the family could better represent the benefit of the donor by collective decision-making. Even so, there are problems related to family consent: Who gives consent on behalf of the family? How about the differences in opinion among family members? I wondered what genetic scientists had to say about family consent in the establishment of genetic databanks.

We asked the genetic scientists the question: In the process of genetic research involving the collection of the donor family's unique genealogical data, do you think that it is necessary to get the family's consent rather than the donor's? Of the respondents, 51.2 per cent gave a negative answer, 47.1 per cent gave a positive one, and 2.7 per cent had no opinion. The survey indicates that the respondents did not hold a consensus on family consent. The results reflect the potential tension between family decision and individual autonomy. We provided the genetic scientists with a list of the major reasons for and against family consent from which the respondents made a selection (Tables 10.3 and 10.4).

The majority of the respondents (82 per cent) emphasise the right of family members in the decision-making process as a major argument for family consent. Twenty-one per cent of them regard family consent as a tradition which has to be followed in genetic research. As far as the arguments against family consent are concerned, the main reason provided by the sampling population is that a donor could make his/her own consent about whether or not to donate his/her blood for genetic research, although most respondents do not deny the role of the family in the decision-making process.

Table10. 3 Arguments for family consent

The major reasons for family consent are: (multi-choice)	*Percentage*
a As the genetic information may be shared by the whole family, family members have the right to join in the decision-making	82%
b A family-based collective decision may better protect individual human subjects	45%
c It is a way of showing respect to the family	44%
d There is a tradition of making decisions together with other family members	21%

Table 10.4 Arguments against family consent

The major reasons against family consent are: (multi-choice)	*Percentage*
a Individual research participants have the right to make a decision autonomously about this	69%
b The donation of a genetic sample may not necessarily harm family members	26%
c Individual decision-making could avoid unnecessary disagreements among family members	19%
d It is not useful to discuss this with other family members	3%

Perceptions of benefit-sharing mechanisms

Ownership

In the case of the Icelandic Health Care Database, the licensee shall not be counted as the owner of the biological samples, but has rights over them, with the limitations laid down by law.[13] The Estonian Genome Project Foundation has the right of ownership of the tissue samples.[14] In China, any institution or individual that holds important pedigrees and genetic resources in those regions with a higher prevalence rate of human genetic diseases shall immediately report to the relevant departments, and no institution or individual may sample, collect, trade, or export human genetic resources; these are the regulations set out by the Interim Measures issued by the MoST and MoH in 1998. Although the Interim Measures do not comment directly on the ownership of the genetic samples, in most cases this reporting and registration system on important pedigrees and genetic resources indicates that governments at different levels control the samples. Thus, the samples belong to central public institutions, rather than that they constitute a kind of public good or are the property of the donors themselves.

To identify our sampling population's attitudes to the ownership of the collected genetic samples/data, we created the following scenario. Suppose there is a big genetic sample collecting project related to hypertension, sponsored by the MoST and carried out by a prestigious Chinese public institute along with a top public hospital. Who should be the owner of the samples? The result of our multi-choice questionnaire indicated that 40.7 per cent of the respondents agreed that the central government should be the owner; 37.2 per cent argued that the samples/data are global public goods; 20.3 per cent thought that the samples/data should be shared by the hospitals and institutions that collect and store the samples; and 7.5 per cent said that the samples belong to the donors themselves. The result shows that the surveyed genetic scientists and ERC members held a diversity of opinion regarding ownership.

In this scenario almost half of the respondents expressed the view that the genetic samples/data should belong to the central government. This is partly because the current genetic databases are state-owned. The practical question here is, however, who has the right to represent the central government? It goes without saying that the MoST and MoH will play a key role in sponsoring and monitoring the genetic databases. Appointed qualified hospitals and institutions would have the right to store and utilise the samples they collected, within the limitations laid down by the Interim Measures, which claim that the Chinese institutions that collected the samples shall have priority access to information about the human genetic resources. Any transfer of important pedigrees and genetic resources and the relevant data to other institutions shall be prohibited unless permission is obtained from the authorities. Even though it does not state that institutions are the owners of the genetic materials, according to the Interim Measures, researchers may eventually acquire intellectual property rights or commercialise products.

Still, about one-third of the respondents regarded the samples/data as global public goods. The Interim Measures also claim that no foreign collaborating institution that has access to the above mentioned information may publicise, publish, and apply for patent right or disclose it by any other means without permission from the authorities. From this it may be concluded that it is only in an abstract sense that the collective human genetic samples form the heritage of humanity as mentioned in United Nations Educational, Scientific and Cultural Organization (UNESCO) and HUGO statements. In reality, the questions of ownership and user-rights of these genetic samples are far from solved.[15]

Commercialisation and conflicts of interest

Although there is little criticism regarding the development of genetic databases in China, many do worry about how to balance the protection and utilisation of Chinese genetic resources. For one thing, in a socialist country such as China, a fundamental assumption is that only state-owned databases can maximise the potential of rare Chinese genetic resources. On the other hand, in the current economic transition period China is going through, the expansion in the construction of genetic repositories could not be sustainable without the positive participation of the private sector. Important questions therefore are: Should a national genetic database be a public or private one? Who will look after the benefits of Chinese donors in any commercial agreements?

There are diverging opinions about the sensitive issue of commercialisation in China. We confronted the genetic scientists and ERC members with the following hypothetical case. One argument is that commercialisation would have a negative effect on the development of Chinese genetic databases; the other is that private genetic databases should be encouraged, because it will contribute to solving the problem of financial support. There is a potential tension between encouragement and discouragement of private databases. In the survey, we posed the following question: Do you think that our central government should set limitations on the commercial investment in the establishment of genetic databases? Of respondents, 37.7 per cent gave a positive answer, 49.3 per cent replied negatively, and 13 per cent did not make a choice.

Of course, commercialisation may generate conflicts of interest, but so does a state monopoly. In the current period, it would be unlikely that China would close the door to private investment in this highly competitive field. The key issue then is how to avoid the adverse consequences of a particular set of conflicts of interest. One of the main areas of conflict of interest in this field of research is rooted in the relation between scientists and pharmaceutical companies, which may give researchers and their institutions a variety of financial benefits.[16] Potential conflict could be avoided if commercial interests were to be disclosed to the ERCs and sample donors via consent forms. This would enable the donors and ERCs to make a sound judgement.

The question now arises of what Chinese scientists think about the disclosure of financial arrangements. The following scenario was put to them: Suppose a public research hospital sponsored by a private pharmaceutical company collects data related to hypertension; should donors be informed? Of the respondents, 59 per cent thought it necessary for the researcher to tell the donor that it is sponsored by a private sector, 37.3 per cent thought it not necessary, while 3.7 per cent had no opinion. The result indicates that a significant proportion of the respondents do not take financial conflicts of interest for granted. In fact, even though the concept of 'class interest' has played a fundamental role in Chinese society throughout the last century, until now the term 'conflict of interest' has not appeared in any dictionary or encyclopaedia in China. It is incumbent upon researchers to explain to the potential donors and/ or their families about the commercial arrangements regarding the development of products with commercial application derived from the research, as well as the commercialisation of the bank itself. However, mere disclosure of commercial interests does not solve the problem entirely. Since conflicts of interest need to be managed properly, there may be a need for other mechanisms for regulation.

Benefits sharing in international cooperation

Scientists in different countries often collaborate on research involving the collection of human genetic samples. The Chinese scientists were confronted with the following scenario. A developed country proposes to provide technology and to train Chinese local researchers. The question raised is: Should China's government supports such a kind of international cooperation? Of the respondents, 37.3 per cent discouraged that kind of cooperation; 39.6 per cent encouraged it; while 23.1 per cent had no opinion. The result shows that the responding scientists of this survey had not developed a clear approach to the proper mechanism for sharing benefits in international cooperation. The Interim Measures, however, do not require closing the door to international cooperation. In fact, the Interim Measures require that the rights and obligations of each party are explicated in order to fully and effectively protect their own respective intellectual property rights. For instance, the right of utilising, transferring, and sharing scientific findings from the collaboration shall be specified in the collaborative contract or agreement signed by both parties.

In consideration of the principle of justice, benefits should be distributed fairly in the process of international cooperation. The unsolved problem is how to make equitable use of genetic data.[17] For example, genetic sample donors in general do not take part in any profit-sharing process, especially in genetic research using public funds. When becoming more aware, donors may start to regard it as unfair when their consent is asked for sampling but no mention is made of genetic ownership and no opportunity is given to share in the commercial benefits derived from the genetic data. HUGO states that even in the absence of profits, immediate health benefits as determined by community needs could be provided if 'profit-making entities dedicate a percentage of their annual net profit to health care infrastructure and/or humanitarian effort'. The Council for International Organizations of Medical Sciences (CIOMS) International Ethical Guidelines for Biomedical Research Involving Human Subjects suggests that externally sponsored research should contribute to national, or local, capacity to conduct biomedical research in the host country.[18] Nevertheless, there must be assurance that the scientific knowledge would be beneficial for the study population.

Actually, benefit sharing could prompt various forms of stimulation to science, such as the diffusion of research results, collaboration with members of the scientific community, and the attribution of licences when the invention resulting from the research is patented. If a research project yields profits, the distribution of benefits could include access to future treatments resulting from the research and donation of a part of the profits to a local humanitarian organisation, financial support for research, and contribution to health technology infrastructures.

Attitudes toward ethical review capacity

Considering the issues discussed above, it is of importance to review protocols involved human genetic sample collecting and usage. In China, however, many ERCs have no extensive policies and procedures regarding human genetic research. In addition, procedures and requirements of ERCs differ from location to location, while there are substantial differences between protocols, which may reflect inconsistent treatment of different types of genetic research. However, on the basis of literature review, we currently only know a little about the attitudes of Chinese scientists and ERC members toward ethical review.

Our survey attempted to identify what Chinese scientists, policy-makers, and ERCs think about the value of ethical review. In this multiple-choice questionnaire, 85.6 per cent of the respondents said it is ethical to protect the rights and benefits of the subjects in genetic sample collecting and usage. There was consensus among the majority of the respondents on the fundamental role of ethical review. In the survey, 66.9 per cent indicated that the ethical review mechanism is meant to protect and exploit Chinese genetic resources rationally. This result may reflect a sense of nationalism. In order to further Chinese national interests, China has taken a few strict measures to safeguard Chinese genetic recourses, which is reflected in the Interim Measures promulgated in 1998.

It is necessary for China to develop its review capacity for genetic sampling projects. Appropriate measures should be taken to improve the ethical review of China's human genetic databases. According to our survey results, 55.4 per cent of scientists and ERC members felt that the data storage process should be standardised for fear of wasting resources, 53.7 per cent of them made an appeal for the state to facilitate the sharing of benefits in international cooperation, 38.8 per cent asked for measures to protect the welfare of donors, 38.1 per cent indicated a need to clarify the ownership of samples and data, while only

21.3 per cent agreed that China should take steps to encourage the free flow of genetic information. It is assumed that the fundamental role of the ERCs is to protect the benefits and rights of human subjects. However, not all respondents took it seriously in the survey. As the Wellcome Trust and MRC report shows, subjects are more likely to participate in a project if sample collection and storage is overseen by an independent body, ideally including a known public figure and members of the study community.[19]

Conclusion and discussion

Our survey on ethical awareness among scientists and ERC members of human genetic databases provides a complex picture of respondents' perceptions and attitudes in general. There was clear disagreement on issues such as family consent and benefit-sharing mechanisms. The sample population in general lacks awareness of the ethics debate in the international community. The government's concern with ethical issues in the creation of genetic databases is also weak. No effective measures were taken in research supervision with regard to ethics. Therefore, appropriate guidelines of ethics review for research involving human genetic samples/data collection and usage is urgently needed. The purpose of the proposed guidelines would be to protect the research participants, to clarify the ethical principles in biomedical research on human subjects, to provide effective ways of ethical supervision, to establish ERCs, to carry out ethics review procedures for the research protocols, and to implement informed consent.

On the issue of disagreement between donors and other family members, genetic samplers should obtain a mutual decision between donors and family members. If the potential donor's opinion conflicts with the family, the investigators may choose to give up for fear of interfering with that family's affairs. This argument reflects the fact that in a community with a paternalistic tradition, individual autonomy is easily dismissed in medicine. Should donors be required, or at least be encouraged, to obtain permission/assent from their family members? The answer may depend on whether results are anticipated that might be returned to respondents and family members. Although, in my view, it is advisable to make decisions with the help of family members, individual preference should still be considered seriously, and incorporated into family decision-making.

Despite the increasing number and expansion of genetic databases, no clear guidelines or regulations are available to regulate benefit sharing. The Interim Measures statements on this topic are limited to prohibiting protocols that involve unfair arrangements of ownership and intellectual property. Unfortunately, the Interim Measures do not provide clear criteria for such unfairness. To remedy this, Chinese policy-makers could develop a kind of genetic sample transfer agreement specifying collaborators' rights and responsibilities with respect to the collection of samples. The key point is that, before both sides agree on the Material Transfer Agreement (MTA), it is a precondition to set up several basic criteria for sharing benefits among different stakeholders.

The result obtained from the survey among scientists and regulators is a good start for encouraging open discussion among stakeholders, including scientists, policy-makers, donors, media, and so on. For many reasons public debate is needed. First, the large scope of genetic sample collections involves a great number of patients and their families. Without their interest and positive participation it is difficult to sustain any kind of such research. Second, the final goal of human genetic databases is to benefit the country, the local communities, and the donors in the long run, so public discussion would contribute to clarifying the objective and to overseeing the procedure followed in the creation of genetic databases. Finally, there is a wide gap between the scientific knowledge that we can obtain from establishing genetic databases and the knowledge we have on the possible consequences of applying this scientific knowledge in the long run.

Therefore, further research still needs to be done by social scientists or other researchers of society on the ethical issues related to genetic databases. For example, a comparative survey should be conducted into the views of Chinese donors, so as to find out about the donors' own attitudes and perceptions, knowledge of which we are currently sorely lacking.

Acknowledgements

This chapter is part of the main finding of a research project funded by the Program on Ethical Issues in International Health of Harvard School of Public Health led by Dr Richard Cash. I would like to thank Professor Daniel Wikler, Professor Qiu Renzong, Professor Gail Henderson, and Dr Margaret Sleeboom-Faulkner for providing valuable advice, insights and comments at various stages in the life of this chapter.

Notes

1 Ghen, X., Zhang, Y., Wang, J.M., Huang, Y., *et al.* (2003) (*Zhong Hua Min Zu Ji Yin Zu DuoYangXIng de GPECG-A Shu Ju Ku*) 'GPCEG-A database for genomic polymorphism of Chinese ethnic groups' (*Yichuan Xuebao*) *Journal of Genetics*, 30(6):509–14.

2 Liu, Junxiang (1999) *Dui Bao Hu He Shi Yong Zhong Guo Ren Yi Chuan Zi Yuan de Fan Si*, 'Reflection on protection and usage of Chinese human genetic resources' (*Yixue yu Shehui*) *Medicine and Society*, 12:337–9.

3 The other sampling institutions include: the Medical School, Zhejiang University; the Medical Center of Peking University; Xiangya Medical School, Zhongnan University; the Medical School of Suzhou University.

4 Gulcher, Jeffrey R. and Stefansson, Kari (2000) 'The Icelandic Healthcare Database and informed consent', *The New England Journal of Medicine*, 342:24.

5 Ministry of Health and Social Security. Act on Biobanks, No. 110/2000, (2000), http://www.stjr.is/interpro/htr/htr.nsf/pages/Act-biobanks, (date accessed: 16 April 2002), s. 7.

6 Xiao, Ze Fen, Yang, Zong Yi and Liang, Jun *et al.* (2003) 'Value of radiotherapy after radial surgery for esophageal carcinoma: a report of 495 patients', *Annals Thoracic Surgery*, 75:331–6.
7 Lu, Yuan (2004) 'Clinical trials in China: protection of subjects' rights and interests', *The Journal of Clinical Ethics*, Spring, 15:1.
8 The author interviewed medical geneticist, Professor Huang Shangzhi, in his laboratory at Peking Union Medical College on 20 March 2006. He mentioned that oral consent is not uncommon, especially when illiterate human subjects are involved.
9 National Bioethics Advisory Commission. Research involving human biological materials: Ethical issues and policy guidance. Rockville, MD, August 1999 (www. bioethics.gov).
10 Autogen Limited, Ethics Policy for Genetics Research Involving the Use of Biological Materials Collected from the People of Tonga, http://www.autogenlim ited.com.au/f_ethics_p2-b.html (date accessed: 12 February 2002,), s. B (5).
11 Medical Research Council (MRC), the Wellcome Trust and the Department of Health, Protocol for the UK Biobank: A study of genes, environment and health, 14 February 2002, http://www.ukbiobank.ac.uk/protocol.htm, p. 31.
12 HUGO, Statement on the Principled Conduct of Genetics Research, 1996, http:// www.hugointernational.org/hugo/conduct.htm (date accessed July 25, 2002).
13 Ministry of Health and Social Security, Act on Biobanks No. 110/2000, (2000), http://www.stjr.is/interpro/htr/htr.nsf/pages/Act-biobanks, (date accessed: 16 April 2002), 10.
14 Estonian Genome Foundation, 'Gene Donor Consent Form', http://www.geeni varamu.ee/mp3/Geenidoonorinousolek- ingl.doc (date accessed: 10 May 2002), 3.
15 Knoppers BM, Fecteau C. (2003) 'Human genomic databases: a global public good?' *European Journal of Health Law*, 10(1):27–41.
16 Qiu Renzong (2004) 'Conflict of interests in research ethics: a Chinese perspective', *The Journal of Clinical Ethics*, 15:1.
17 Sulston J. (2003) 'Beyond release: the equitable use of genomic information', *Lancet*, 362(9381):400–2.
18 Council for International Organizations of Medical Sciences (CIOMS): International Ethical Guidelines for Biomedical Research Involving Human Subjects. Switzerland: CIOMS, 2002, (Accessed December 2004). http://www.cioms.ch/ frame_guidelines_nov_2002.htm
19 Wellcome Trust and MRC (2000) *Report: Public perceptions of the collection of human biological samples*. Wellcome Trust & MRC.

Index